PRIVATE ACTION/PUBLIC GOOD:
*Maryland's Nonprofit Sector
in a Time of Change*

PRIVATE ACTION/PUBLIC GOOD:
Maryland's Nonprofit Sector in a Time of Change

Lester M. Salamon

with the assistance of
Peter Berns, Amy Coates Madsen
Wojciech Sokolowski, and Stefan Toepler

Maryland Association of Nonprofit Organizations

Institute for Policy Studies Johns Hopkins University

Maryland Association of Nonprofit Organizations
190 West Ostend Street, Suite 201
Baltimore, MD 21230

Copyright © 1997 Maryland Association of Nonprofit Organizations

All rights reserved. Except for short quotations, no part of this book may be reproduced or utilized in any form or by any means, electronic or mechanical, including photocopying, recording, or by information storage or retrieval system, without written permission from Maryland Association of Nonprofit Organizations or Lester M. Salamon.

Private Action/Public Good/ by Lester M. Salamon

ISBN 0-9660113-0-9

Library of Congress Catalog Card Number: 97-74707

TABLE OF CONTENTS

List of Tables vii
List of Figures viii
Preface xi
Acknowledgments xiii
Executive Summary xv

Chapter One Introduction 1
    Purpose of the Report 2
    Focus: Public Benefit Service Organizations 2
    Geographic Focus 3
    Methodology 3
    Structure of the Report 5
    Summary 5

Chapter Two A Major Economic Force 7
    The Overall Size of the Maryland Nonprofit Sector 8
    Types of Nonprofit Organizations 12
    Regional Distribution 19
    Conclusion 26

Chapter Three A Closer Look: The Structure and Focus of Maryland's Nonprofit Sector ..27
    What Maryland Nonprofit Organizations Do 27
    Agency Age 35
    Whom Does the Maryland Nonprofit Sector Serve? 38
    Conclusion 44

Chapter Four Nonprofit Finances 45
    Overview of Maryland Nonprofit Finances 45
    Patterns of Nonprofit Finance 49
    Bringing in Volunteers 54
    Conclusions 55

Chapter Five Challenges and Response 57
    The Challenges 57
    Nonprofit Responses 65
    Conclusion 71

Chapter Six Ethics and Accountability 73
    Standards of Ethical Management 73
    The Board of Directors: The Pivot of Nonprofit Management 74
    Achieving the Mission: The Nonprofit Sector's Bottom Line 78
    Living in a Glass House: The Duty to Disclose Information to the Public 80
    Promoting Public Trust in Maryland's Nonprofit Sector 81
    Conclusion 82

Chapter Seven Nonprofits in Maryland's Regions 83
    The Baltimore Region: Traditional Capital of Maryland's Nonprofit Sector 83
    The Washington Suburbs: Building a Civic Infrastructure 86
    Western Maryland and the Eastern Shore 89
    Conclusion 91

Chapter Eight Conclusions and Recommendations..93
    The Need for Renewal: A Blueprint for Action 94
    1. Engaging Opinion Leaders: A Maryland Civil Society Commission 94
    2. Monitoring the Health of the State's Nonprofit Sector 95
    3. Boosting Private Giving and Buttressing the State's Private Philanthropic Base 96
    4. Building Organizational Capacity..98
    5. Promoting Public Confidence..100
    Conclusion 101

Appendix A: Maryland Nonprofit Survey Methodology 103
Appendix B: Fields of Activity of Maryland Nonprofit Agencies 112

About the Author 117

LIST OF TABLES

Table 1.1 Maryland Resident Population by Region, 1995 3

Table 2.1 Paid and Volunteer Employment in Maryland's Nonprofit Sector
as a Share of Total Employment, 1996 8

Table 2.2 Wages and Operating Expenditures of
Maryland Nonprofit Organizations, 1996 10

Table 2.3 Growth of Maryland Nonprofit Employment, 1989-1996 10

Table 2.4 Nonprofit Sector Employment, by Industry, Maryland vs. U.S. Average 14

Table 2.5 Change in Nonprofit Share of Private Employment, by Field, 1996 vs. 1989 18

Table 2.6 Share of Nonprofit Employment Growth in Maryland, by Region, 1989-1996 22

Table 2.7 Composition of the Nonprofit Sector, by Industry, by Region, 1996 23

Table 2.8 Relative Share of Nonprofit Growth, 1989-1996, by Field, by Region 25

Table 3.1 Fields of Activity of Maryland Nonprofit Agencies, 1996 28

Table 3.2 Average Agency Size, by Type of Agency, 1994 31

Table 3.3 Distribution of Maryland Nonprofit Expenditures
and Agencies, by Field, 1994 32

Table 3.4 Distribution of Nonprofit Expenditures, by Field, Maryland and U.S., 1994 33

Table 3.5 Distribution of Maryland Nonprofit Agencies, Employees,
and Volunteers, by Size of Agency, 1994 34

Table 3.6 Distribution of Maryland Nonprofit Agency Activity,
by Field, with Volunteers 35

Table 3.7 Share of Maryland Nonprofit Agencies and Expenditures,
by Age of Agency, 1994 38

Table 3.8 Geographic Focus of Maryland Nonprofit Agencies, by Age of Agencies 40

Table 4.1 Sources of Support for Maryland Nonprofit Organizations.....................................47

Table 4.2 Sources of Income of Maryland Nonprofit Agencies, by Client Focus 53

Table 5.1 Recent Changes in Operating Revenues of Maryland Nonprofit Agencies,
1989-94, by Size of Agency and Source of Income 60

Table 5.2 Concerns about For-Profit Competition Among Maryland
Nonprofit Agencies, by Size of Agency...64

Table 5.3 Anticipated Changes in Revenues of Maryland
Nonprofit Organizations, by Source, 1996-98 70

Table 6.1 Share of Maryland Nonprofit Organizations Reporting Various
Disclosure Practices, by Size of Agency..80

Table 7.1 Baltimore Area Nonprofits: A Profile 84

Table 7.2 The Nonprofit Sector in the Washington Suburbs of Maryland: A Profile 87

Table 7.3 Nonprofit Organizations in Western Maryland
and the Eastern Shore: A Profile 90

LIST OF FIGURES

Figure 2.1 Employment in Maryland Nonprofit Sector and Selected Industries, 1996 9

Figure 2.2 Changes in the Number of Jobs in the Nonprofit Sector and Other Industries in Maryland, 1989-1996 11

Figure 2.3 Average Annual Growth Rates of Nonprofit Employment and Total Maryland Employment, 1989-1992, 1992-1995, 1995-1996 12

Figure 2.4 Employment in Maryland Nonprofit Organizations, by Field, 1996 13

Figure 2.5 Nonprofit Share of Maryland Employment, by Field, 1996 15

Figure 2.6 Changes in Nonprofit Employment in Maryland, by Field, 1989-96 15

Figure 2.7 Share of Nonprofit Employment Growth in Maryland, 1989-96, by Field 17

Figure 2.8 Distribution of Maryland Nonprofit Employment, by Region, 1996 19

Figure 2.9 Nonprofit Share of Total Employment in Maryland, by Region, 1996 20

Figure 2.10 Growth of Nonprofit and Total Employment in Maryland, by Region, 1989-96 21

Figure 3.1 Distribution of Maryland Nonprofit Organizations, by Primary Field, 1996 29

Figure 3.2 The Largest 7 Percent of Maryland Nonprofits Account for 93 Percent of Expenditures, 1994 30

Figure 3.3 Distribution of Nonprofit Expenditures, by Field, 1994 32

Figure 3.4 Distribution of Maryland Nonprofit Activity, Including Volunteers, by Major Field, 1994 34

Figure 3.5 Distribution of Maryland Nonprofit Agencies by Year of Establishment 36

Figure 3.6 Percent of Maryland Nonprofit Agencies Formed Since 1980, by Field 37

Figure 3.7 Geographic Service Area of Maryland Nonprofits, 1996 39

Figure 3.8 Client Focus of Maryland Nonprofit Agencies 41

Figure 3.9 Characteristics of Primarily Poor Serving Maryland Nonprofit Agencies 42

Figure 4.1 Major Revenue Sources of Maryland Nonprofit Agencies, 1994 46

Figure 4.2 Sources of Income of Maryland Nonprofit Agencies, by Field, Size and Age of Agencies, 1994 50

Figure 4.3 Percent of Maryland Nonprofit Agencies, with Indicated Principal Sources of Income, 1994 51

Figure 4.4 Sources of Income of Maryland Nonprofit Agencies, by Region, 1994 54

Figure 4.5 Sources of Income of Maryland Nonprofit Agencies, including and excluding Volunteer Input 55

Figure 4.6 Sources of Income of Maryland Nonprofit Agencies, with Volunteers, by Type of Agency, 1994 56

Figure 5.1 Changes in Demand for Nonprofit Services in Maryland, 1994-96 58
Figure 5.2 Share of Maryland Nonprofit Agencies with Real Income Growth,
      1989-94, by Size and Type of Agency 59
Figure 5.3 Share of Maryland Nonprofit Agencies with Real Income Growth,
      1989-94, by Source 60
Figure 5.4 Share of Maryland Nonprofit Organizations Reporting Staffing Problems 61
Figure 5.5 Perceptions of Public Attitudes among Maryland Nonprofits................................63
Figure 5.6 Share of Agencies Reporting Increased Market Pressures, by Size of Agency 66
Figure 5.7 Share of Maryland Nonprofit Agencies Reporting Recent Changes
      in Service Levels and Client Focus..66
Figure 5.8 New Nonprofit Management Practices of Maryland Nonprofit Agencies 68
Figure 5.9 New Fundraising Practices of Maryland Nonprofit Agencies 69
Figure 6.1 Structure and Operations of Maryland Nonprofit Boards:
      Expectations vs. Performance...75
Figure 6.2 Actual vs. Desired Board Roles in Managing Nonprofit Organizations..................78
Figure 6.3 Management and Disclosure Practices of Maryland Nonprofit Organizations:
      Expectations vs. Performance...79
Figure 6.4 Maryland Nonprofit Agency Proposals for Promoting Public Trust 81

PREFACE

In his 1990 report, *More than Just Charity: The Baltimore Area Nonprofit Sector in a Time of Change (1990),* Dr. Lester M. Salamon of the Johns Hopkins University Institute for Policy Studies startled the local community with his revelations about the size, scope, structure, and financing of Baltimore's nonprofit sector. That report demonstrated that the nonprofit sector in the Baltimore metropolitan region was a major player in the economy of the area, employing more people and pumping more money into the local economy than had ever been imagined or expected. In the report, Dr. Salamon also described the significant contribution nonprofits make to the quality of life in the region, providing programs and services that touch the lives of more than 2.3 million area residents.

More Than Just Charity's findings and recommendations were of more than academic interest in the nonprofit and philanthropic community--they were a catalyst for change. The report offered a series of recommendations to strengthen the region's nonprofit sector and many of these have since been implemented. In fact, one of those recommendations led to the creation of the Maryland Association of Nonprofit Organizations (Maryland Nonprofits).

In the intervening years, Maryland Nonprofits has worked to organize Maryland's nonprofit sector, provide training and technical assistance, create cooperative buying programs for nonprofits, and educate political leaders, government officials, business and community leaders, and the general public about the important role of the state's nonprofit community.

With the publication of the present report, Maryland Nonprofits takes a giant step toward this latter goal. *Private Action/Public Good: Maryland's Nonprofit Sector in a Time of Change* is the first effort that has ever been undertaken to provide a comprehensive picture of the state's entire nonprofit sector.

We are pleased and privileged to have Lester M. Salamon serve as principal investigator and author of this important report. The nonprofit sector in Maryland is extraordinarily fortunate to have Dr. Salamon as part of our community. Recognized as the preeminent authority on the nonprofit sector, nationally and internationally, Dr. Salamon brings a wealth of knowledge, insight and wisdom to enrich our understanding of Maryland's nonprofit sector.

Just as the 1990 report startled readers, *Private Action/Public Good: Maryland's Nonprofit Sector in a Time of Change* is sure to astound as well. The report describes a nonprofit sector that has grown substantially over the past six years and that plays a more significant role in the economy than anyone expected. It also describes a set of organizations that reaches into every corner of the state, from the Eastern Shore to the mountains of Western Maryland; and that touches the state's residents in a hundred different ways.

Of great concern, however, is the report's description of the vulnerability of the state's nonprofit sector. Dr. Salamon describes how Maryland's nonprofit organizations are struggling to meet ever increasing demands for service in the face of diminishing government support, lackluster charitable giving, and increasing public skepticism. Fortunately, Dr. Salamon also offers us some solutions. The report's "Blueprint for Action" provides a five-point plan for the renewal of Maryland's nonprofit sector that is both straightforward and achievable.

Private Action/Public Good should be studied carefully by government officials, nonprofit staff, board members, business and community leaders, and anyone who is interested in the success of Maryland's nonprofit sector and the quality of life in our state. The report's recommendations should provide the starting point for a statewide and community-wide effort to strengthen and renew this important segment of Maryland society. That effort needs to start today.

Peter V. Berns
Executive Director
Maryland Association of Nonprofit Organizations

ACKNOWLEDGMENTS

This report represents an attempt to place on the economic, political, and social map of the State of Maryland a set of institutions on which the state's citizens rely every day but of which they are largely unaware-- the thousands of day care centers, hospitals, higher education institutions, social service agencies, employment and training facilities, theaters, museums, symphonies, and countless others that comprise the state's private nonprofit sector.

Although it was my privilege to carry out this study, the inspiration for it, and much of the hard work that made it possible, came largely from Peter Berns, the able and energetic executive director of the Maryland Association of Nonprofit Organizations (Maryland Nonprofits), and Amy Coates Madsen, his thorough special assistant. It was Peter who first conceived the idea of following up on an earlier study I had done of the Baltimore area nonprofit sector. It was Peter who convinced me to take on this task. It was Peter who finally managed to secure the funding needed to support the effort. And it was Peter who, along with Amy, worked with me at every stage to design the survey, determine the sampling strategy, field the survey, convert the results into computer-readable form, and then help prepare the resulting manuscript for publication. The "assistance" mentioned on the title page thus understates by far the contribution that Peter and Amy made to this report. Both were full-fledged collaborators on this project without whose help the report would not have been possible.

In addition to Peter and Amy, this project received invaluable assistance from Drs. Wojciech Sokolowski and Stefan Toepler of the Johns Hopkins Institute for Policy Studies, who carried out the data processing and graphics development work on the survey and employment data, respectively. In the case of the survey in particular this required hours of rather complex data analysis and re-analysis and often intricate detective work to ferret out various glitches and anomalies in the data. Thanks are also due to Mr. Patrick Arnold, Director of the Division of Labor Market Analysis and Information in the Maryland Department of Labor, Licensing and Regulation, and to the Honorable Eugene A. Conti, Jr. Secretary of that Department, for the tremendous service they did in pulling nonprofit firms out of the regular employment data that the Department collects from employers throughout the state; to Beverly Valcovic and Nikki Baines of the Maryland Secretary of State's office and to the Honorable John Willis, Secretary of State for helpful information and assistance; to Tom Brown of the Johns Hopkins Department of Sociology, who assisted with the initial sampling; to Marcy Shackelford and Jacquelyn Perry of the Johns Hopkins Institute for Policy Studies, who made major contributions to the production of the manuscript; and to the hundreds of Maryland nonprofit organization officials who took time from their busy schedules to complete the survey on which much of this report is based.

Finally, I want to express my appreciation to the funders of this project: the Morris Goldseker Foundation, the Nonprofit Sector Research Fund of the Aspen Institute, the Eugene and Agnes E. Meyer Foundation, the Abell Foundation, the National Society of Fundraising Executives-Maryland Chapter, the Association of Baltimore Area Grantmakers, and the Chesapeake Planned Giving Council.

Despite the help I have received, any conclusions or opinions expressed in this report are my own and are not necessarily shared by those who assisted with the effort, by the institutions with which we are affiliated, or by those that supported the work. As with any project of this sort, the final task of unraveling a complex body of data of the kind assembled here and making sense of it involves as much art as science. Judgments inevitably have to be made about which findings are significant and which not, which points deserve highlighting and which are trivial, which results should be questioned and which believed. In the present study, those judgments fell to me to make. Ultimately, therefore, responsibility for any errors of fact or interpretation that remain is mine alone, and I accept it gladly.

Lester M. Salamon
Arnold, Maryland
August 22, 1997

Private Action/Public Good: Maryland's Nonprofit Sector in a Time of Change

Lester M. Salamon
Executive Summary *

Introduction

Beyond the institutions of government and private business so familiar to citizens of Maryland lies a vast collection of other organizations that play a crucial role in community life but that are largely unknown to most residents, unreported in the media, and unexamined by either policy makers or researchers. This collection of organizations is known variously as the "nonprofit," the "voluntary," or the "charitable" sector and it includes thousands of private day care centers, adoption agencies, family counseling programs, employment and training facilities, neighborhood organizations, nursing homes, hospitals, colleges, universities, schools, self-help groups, museums, art galleries, theaters, and others.

The purpose of this report is to fill the vast gap in knowledge that has long existed about these organizations in Maryland and thus to provide a better foundation for appraising their current role and future prospects.

To do so, the report draws on three principal sources of data:

- A new body of employment data generated specially for this project by the **Maryland Department of Labor, Licensing, and Regulation (DLLR)**.

- A **survey of nonprofit public-benefit organizations** conducted by this author in cooperation with the Maryland Association of Nonprofit Organizations.

- The **U.S. Census of Service Industries and related data** used to convert employment estimates into estimates of operating expenditures.

The focus of the report is that portion of the nonprofit sector that serves essentially public purposes. Such organizations enjoy exemption from federal income taxation under two of the more than 25 provisions of the U.S. tax code that provide for such exemption--namely, sections 501(c)(3) and 501(c)(4).[1]

Major Findings

Twelve major conclusions emerge from this work, as outlined below.

1. Maryland's Nonprofit Sector is a Major Economic Force.

- **Employment.** Nonprofit organizations employed just over 185,000 workers, or one out of every 12 workers, in Maryland as of the end of 1996.

---

*Full copies of this report are available from the Maryland Association of Nonprofit Organizations, 190 Ostend Street, Suite 201, Baltimore, MD 21230, for a cost of $35.00.

[1]Section 501(c)(3) grants tax exemption to organizations that are "organized and operated exclusively for religious, charitable, scientific, testing for public safety, literary, or educational purposes, or to foster national or international amateur sports competition or for the prevention of cruelty to children or animals..." Section 501(c)(4) grants exemption to "civic leagues or organizations not organized for profit but operated exclusively for the promotion of social welfare."

- More people are therefore employed in the nonprofit sector in Maryland than in *all* of the state's manufacturing industries, *all* of its construction businesses, *all* of its finance and real estate businesses, *all* of its transportation and communications businesses, and both the federal government and the state government (See Figure 2.1).

- With volunteer effort included, the nonprofit sector accounts for an even larger 11.4 percent of the state's total workforce.

- **Wages and expenditures.** The paid employment of nonprofit organizations alone accounted for over $5 billion in wages in 1996, or close to 8 percent of the state's total wages. The overall expenditures totaled close to $13 billion, or the equivalent of almost $2,600 per Maryland resident.

- Three fields absorb the majority of nonprofit employment in Maryland:

  - **Health**, which accounts for over 50 percent of the nonprofit employment;

  - **Education and research,** which accounts for 21 percent; and

  - **Social services,** which accounts for 19 percent (See Figure 2.4).

2. MARYLAND'S NONPROFIT SECTOR HAS BEEN A MAJOR CONTRIBUTOR TO STATE EMPLOYMENT GROWTH.

- Nonprofit organizations accounted for half of the net new jobs that Maryland generated between 1989 and 1996--35,000 new jobs compared to 67,274 new jobs for all the state's industries. Compared to the 35,000 new jobs that the nonprofit sector generated, Maryland's manufacturing firms lost 30,175 jobs, its construction industry lost 26,228 jobs, and its wholesale trade industry lost 4,905 jobs (See Figure 2.2).

More people work in nonprofit organizations in Maryland than in manufacturing.

- Nonprofit job growth was particularly robust in the fields of culture and arts (+87 percent), social services (+49 percent), and health (+24 percent). By contrast, it was much more restrained in the field of education (+11 percent). In virtually every field, however, nonprofits added jobs at a rate substantially greater than that of the overall state economy.

- The growth rate of nonprofit employment has slowed considerably in the past couple of years, however, dropping from an average annual growth rate of 4.9 percent between 1989 and 1992 to only 1.0 percent between 1995 and 1996.

Over half of the total job growth in Maryland between 1989 and 1996 came from the nonprofit sector.

- Despite often substantial growth, nonprofit organizations appear to be losing market share to for-profit firms in the larger fields of nonprofit action--i.e. health, education, and social services. This is particularly striking in the field of social services, where nonprofit organizations suffered a 6 percent decline in their market position relative to for-profit companies between 1989 and 1996.

3. NONPROFIT ORGANIZATIONS ARE PRESENT IN EVERY REGION OF THE STATE.

- Even in Western Maryland and the Eastern Shore, nonprofit organizations account for 6 to 8 percent of total employment, thus outdistancing construction, transportation, wholesale trade, and real estate and insurance as sources of employment.

- Although the nonprofit sector is present in all portions of the state, like the state's population more generally its "capital" is clearly Baltimore City and the Baltimore region. Thirty-nine percent of all Maryland nonprofit employees work in Baltimore City alone and another 25 percent in the Baltimore metropolitan area outside of the city. In fact, 18 percent of the city's jobs--or one out of five--are in the nonprofit sector.

- Like the state's population, however, nonprofit employment has been *decentralizing*, with the greatest growth in the outlying regions. Thus nonprofit employment grew by 9 percent in Baltimore City between 1989 and 1996, but by 22 percent in Western Maryland, 36 percent on the Eastern Shore, 38 percent in the Washington suburbs, and 40 percent in the Baltimore suburbs. In the process it outdistanced overall employment growth in every region (See Figure 2.10).

> *Even in Western Maryland and the Eastern Shore, nonprofit organizations outdistance construction, transportation, wholesale trade, and real estate and insurance as sources of employment.*

4. THE NONPROFIT SECTOR IS EXTREMELY DIVERSE, TOUCHING VIRTUALLY EVERY MAN, WOMAN, AND CHILD IN THE STATE.

- The more than 12,000 nonprofit organizations operating in Maryland affect virtually every aspect of community life, from health and education to culture and recreation.

- By far the most common field of nonprofit action is *social services*. Just over half of all Maryland agencies reported some involvement in this field (See Table 3.1).

- Just behind social services as a focus of nonprofit action is *education*, which engages nearly half of Maryland nonprofit organizations. Included here are actual educational institutions as well as parent-teacher associations and information centers of various sorts.

- The third most commonly cited activity of Maryland nonprofit organizations, significantly, is *advocacy*. Just over 30 percent of all agencies reported some involvement in advocacy, civil rights, or legal rights. In addition, 29 percent indicated some involvement in community development, which is very similar to advocacy.

- When measured in terms of "primary service field," i.e., the field that absorbs the majority of agency expenditures, three fields--*culture, arts, and recreation; social ser-*

> *There are few areas of community life in which nonprofit organizations are not making a contribution in Maryland.*

vices; and *education*--clearly attract the majority of Maryland nonprofit agencies. Taken together, these three components account for over half (56 percent) of all nonprofit agencies in the state.

- Maryland nonprofit agencies also differ considerably by size:

  - Two-thirds of Maryland nonprofit organizations have total expenditures of less than $25,000. Taken together, however, these agencies account for only 1 percent of total nonprofit expenditures in the state (See Figure 3.2).

  - At the opposite extreme, 7 percent of the agencies have expenditures of $1 million or more. However, these agencies account for 93 percent of all expenditures.

  - Smaller agencies make more extensive use of volunteers than they do of paid staff (13 percent vs. 1 percent). However, the 7 percent of agencies in the largest size category still absorbed a disproportionate share (43 percent) of the volunteers.

The top 7 percent of Maryland nonprofit organizations account for 93 percent of total nonprofit expenditures.

5. THE MARYLAND NONPROFIT SECTOR IS DYNAMIC.

- Almost two-thirds of the organizations in existence at the time we conducted our survey in 1996 had been created since 1971, and over 40 percent since 1981. This suggests the critical role that the nonprofit sector performs as a mechanism for surfacing and responding to new social concerns and thus as a significant source of social vitality. It may also reflect the movement of nonprofit organizations to the suburbs identified above.

Almost two-thirds of the organizations in existence in 1996 had been created since 1981.

- Especially large shares of newer agencies are evident in the employment and training field, in social services, in community development and environmental protection, and in the "other" category. By contrast, much higher proportions of multipurpose, advocacy, and education organizations were formed prior to 1961. Included here are the large family service agencies that have become so pivotal a part of the human service scene in the state (See Figure 3.6).

- While the younger agencies are far more numerous than the older ones, they are also much smaller. Thus, the 26 percent of all agencies formed prior to the 1960s account for a striking 84 percent of the expenditures.

6. MARYLAND NONPROFIT ORGANIZATIONS SERVE A BROAD CROSS-SECTION OF THE STATE'S CITIZENS.

- Maryland nonprofit organizations perform a variety of social roles. Many serve a community-building function, linking individuals to the communities where they live. Reflecting this, the largest single group of agencies, accounting for 27 percent of the total, report a neighborhood focus, and another 26 percent operate primarily at the city or county level.

- Another quarter of Maryland's nonprofit organizations serve a metropolitan or statewide clientele. They therefore provide a mechanism for addressing problems that span local governmental jurisdictions.

- Despite the conventional notion that nonprofit organizations primarily serve the poor, most agencies serve a diverse clientele. The poor thus comprise the majority of the clients of only 16 percent of the agencies. And only 26 percent of the agencies reported that the poor comprise as many as 10 percent of their clients. Clearly, Maryland's nonprofit sector is by no means primarily "charitable" in the dictionary sense of the term (See Figure 3.8).

Maryland's nonprofit sector is not primarily "charitable" in the narrow sense of the word.

- The agencies most likely to focus primarily on the poor are large agencies, younger agencies, agencies in the fields of health and mental health, and agencies located in Baltimore City (See Figure 3.9). In no case, however, does a majority of the agencies report focusing primarily on the poor.

7. PRIVATE CHARITABLE GIVING PLAYS A MUCH SMALLER ROLE IN THE FINANCING OF NONPROFIT ACTIVITY THAN IS WIDELY BELIEVED.

- More than three-fourths (78 percent) of Maryland nonprofit agencies report some income from private giving, including giving by individuals, foundations and corporations; yet private giving accounts for only 4 percent of total nonprofit sector income in the state (See Figure 4.1).

- This means that private philanthropy's share of total nonprofit income is *proportionally two and one-half times smaller in Maryland than it is nationally (4 percent vs. 10 percent).*

- Instead of private giving, the major source of support for Maryland nonprofit agencies is *earned income* from fees, related businesses, unrelated businesses, and investments. Such earned income comprises over 50 percent of the total revenue of Maryland nonprofit agencies, most of it from fees and charges, and almost as many agencies receive income from this source (70 percent) as from private giving.

- If earned income is the major source of nonprofit income in Maryland, *government* support is a close second. Federal, state, and local governments provide 44 percent of the income of Maryland nonprofit agencies, compared to 34 percent nationally.

- Of the ten major types of agencies we identified in our survey, five get the preponderance of their income from government support. Included here are mental health, employment and training, social services, health, and multipurpose agencies. (See Figure 4.2).

Private giving comprises less than 5 percent of total nonprofit sector income in Maryland, much smaller than is the case nationally.

- Of the five remaining types of agencies, four get the majority of their support from earned income. Included here are community development, education, culture and recreation, and other agencies. In addition, earned income accounts for more than 49 percent of total income in the huge field of health, where it shares top funding honors with government.

- In only one field--advocacy--does private giving provide the majority of funding, and this is a relatively small, though important, field of nonprofit action.

- Private giving is a far more significant source of income for small agencies and young agencies. For the latter, however, the *largest* source of income is still government. Sixty percent of the income of the youngest agencies, and 54 percent of the income of the next youngest agencies, comes from government. This suggests the important "enabling" function that government performs for nonprofit organizations in Maryland.

Agencies that primarily serve the poor tend to have much higher levels of government support than those that do not.

- Agencies that primarily serve the poor also tend to have much higher levels of government support than those that do not. Indeed, among primarily poor-serving agencies, government accounts for 74 percent of total income.

- With volunteer time included, the share of private giving in total nonprofit revenue more than triples--from 4 percent to 15 percent. At the same time, even with volunteer time included, private giving remains the third most important source of income for Maryland nonprofit agencies, and a distant third at that.

8. MARYLAND NONPROFIT ORGANIZATIONS HAVE BEEN CONFRONTING SERIOUS CHALLENGES IN RECENT YEARS. THESE CHALLENGES TAKE AT LEAST FIVE DIFFERENT FORMS.

- **Increased Demand for Services.** In the first place, the early 1990s was a period of expanding demand for the services that Maryland nonprofit agencies provide. Nearly two-thirds of the agencies (64 percent) reported noticeable increases in the demand for their services during the previous two years, and for nearly 40 percent the increases were substantial (i.e. 10 percent or more) (See Figure 5.1).

- **Constrained Income Growth.** In the face of this growth in demand, nonprofit agencies in Maryland reported only modest growth in income. Although 45 percent of Maryland agencies reported increases in income between 1989 and 1994, for only 14 percent did the reported increase exceed the rate of inflation (See Figure 5.2). This means that 86 percent of Maryland's nonprofit agencies were not able to boost their incomes enough to keep pace with inflation. Virtually all the sources of income contributed to this tepid performance, moreover.

- **Staffing Challenges.** Reflecting these fiscal pressures, Maryland nonprofit organizations encountered a variety of staffing problems. Thus, two-thirds of the agencies reported problems maintaining competitive benefit packages for their employees; and over 60 percent reported increases in the workload per paid employee over the previous two years. Nonprofit agencies thus seem to be demanding more of their employees while finding themselves unable to reward the increased work with higher pay.

- **Volunteer Challenges.** In addition to experiencing problems recruiting and retaining *paid* staff, Maryland nonprofit organizations are also experiencing problems with the volunteer side of their operations. Thus, over 70 percent of the agencies reported difficulties recruiting dependable, qualified volunteers; 62 percent reported difficulties retaining such volunteers once recruited; and close to 70 percent reported problems providing sufficient training for volunteers.

What these data make clear is that the volunteer "solution" is hardly a panacea for the staffing problems confronting Maryland nonprofit organizations. To the contrary, volunteerism itself faces important challenges in the state.

The volunteer "solution" is hardly a panacea for the staffing problems confronting Maryland nonprofit organizations.

- **Public Attitudes.** Added to the fiscal pressures and resulting staffing difficulties they are confronting, Maryland nonprofit organizations are also facing an apparent lack of public support. Over half of the nonprofit executives we surveyed report that they believe that "the public is becoming increasingly distrustful of nonprofit organizations."

- **For-Profit Competition.** Finally, Maryland nonprofit organizations are confronting increased for-profit competition in traditional areas of nonprofit operation.

9. MARYLAND NONPROFITS HAVE RESPONDED TO THE CHALLENGES THEY FACE BY MAKING MANAGEMENT AND FUNDRAISING CHANGES.

- **Protecting the Client Base.** In the face of these pressures, a quarter of all agencies, and half of the large agencies, reported that funding pressures were pushing them toward "becoming less responsive to client needs and more responsive to market forces."

Over half of the nonprofit executives we surveyed report that they believe that "the public is becoming increasingly distrustful of nonprofit organizations."

- In response, some 15-16 percent of all agencies, and 35 percent of the larger organizations, introduced new client fees or increased those already in operation.

- At the same time, agencies appear to be resisting more direct changes that would restrict client access. Thus, only 5 percent tightened their eligibility requirements, only 6 percent reduced the number of services they provide or the number of people served, and only 1 percent reduced their hours of service.

- **Management Changes.** Instead, Maryland nonprofits have turned to a variety of management changes (See Figure 5.8). Thus:

  - Almost half of all agencies are making more extensive use of *volunteers* in their basic program operations;

  - More than 45 percent of all agencies, and 90 percent of the large agencies, have recently instituted new management practices;

  - Other widespread management changes adopted over the past two years involve *increasing user involvement* (28 percent of agencies), *sharing resources with other agencies* (27 percent), *increased staff training* (27 percent), and *reorganizing agency functions* (21 percent).

- **Fundraising Changes.** Perhaps the major strategy Maryland nonprofit organizations are relying on to withstand the pressures they are under is to change their fundraising practices.

  - Nearly two-thirds of Maryland's nonprofit organizations report that they tried at least one new fundraising approach in the previous two years, and three-fourths report that they are planning additional changes over the next two years.

- A third of all agencies applied *to new corporate and foundation donors* in the past two years and half expect to make a try over the next two years.

- About a quarter of the agencies reported trying *special events fundraising, product sales, and new government programs* over the previous two years and between 40 percent and 50 percent plan to turn to them over the next two years.

- Among large agencies, over 60 percent increased their fees over the previous two years.

- **Anticipated Funding Trends.** Maryland nonprofit organizations are rather optimistic about the likelihood that these new fundraising strategies will yield results. Thus, over half (55 percent) expect to boost their revenues by at least 10 percent over the next two years. Given the prevailing patterns of funding, however, there is reason to be somewhat skeptical about these expectations.

10. MARYLAND NONPROFIT ORGANIZATIONS GENERALLY FOLLOW ETHICAL MANAGEMENT PRACTICES, BUT SIGNIFICANT ROOM FOR IMPROVEMENT STILL EXISTS.

- Maryland nonprofit boards appear to meet the basic structural and operational characteristics thought to be most conducive to effective operation. Thus, nine out of ten boards contain 5 or more members and over 90 percent meet at least four times a year (See Figure 6.1).

- Although virtually no nonprofit board members are compensated for their service, a significant minority of agencies (31 percent) reported that their board members use their positions for "personal enrichment," and 9 percent reported that their organizations had purchased goods or services from a board member or a board member's immediate family in the past year.

- Despite this, only 21 percent of the agencies reported having written conflict of interest policies in place. Only among the large nonprofits does a majority (62 percent) have such a written policy in place. For medium and smaller nonprofits, written conflict of interest policies are the exception rather than the rule (10 percent and 33 percent, respectively).

- Most Maryland nonprofit boards are substantially or highly involved in the central management functions of their organizations--determining the organization's mission, developing its strategic plan, selecting its board members.

- However, only about half of the organizations indicated that their boards were substantially or highly involved in reviewing the organization's performance or setting its fundraising strategy; and *less than half* indicated that the boards had this level of involvement in hiring, firing, and evaluating the executive director, setting executive compensation, deciding to purchase goods or services from a board member, or setting policy on fees (See Figure 6.2).

- As stipulated in recommended practice in this field, the vast majority of Maryland nonprofits (85%) report having a written mission statement. However, fewer than half of the organizations reported having systems in place to evaluate the success of *any* of their programs,

Only one in five Maryland nonprofit agencies has a written conflict of interest policy in place.

and less than 20 percent reported having systems in place to evaluate *all* of their programs.

- Similarly, while virtually all respondents acknowledged the importance of disclosure, only three-quarters reported that they regularly circulate program information; only 54 percent publish an annual report; and less than half (49 percent) regularly publish financial statements.

- In the face of the gaps in accountability noted above, few nonprofit leaders (15 percent) believe that there is a need for additional government regulation. However, a plurality (43%) believe that "there should be more vigorous enforcement of existing laws and regulations governing charitable solicitations", 57 percent favor more effective self-regulation by nonprofits and 59 percent agree that the nonprofit sector should develop a strong code of conduct for nonprofit management practices.

Fewer than half of the organizations reported having systems in place to evaluate the success of any of their programs.

11. THE SCOPE AND STRUCTURE OF THE NONPROFIT SECTOR DIFFERS CONSIDERABLY IN DIFFERENT PARTS OF THE STATE.

- **Baltimore area.** The Baltimore region, historically the hub of Maryland's economy and its political power, is also the capital of its nonprofit sector.

  - This region houses half of the state's nonprofit institutions and accounts for close to two-thirds of its nonprofit employment.

  - Close to 11 percent of all employment in the Baltimore area--one out of every 9 jobs--is in the nonprofit sector, and in Baltimore City alone this figure reaches 18 percent.

  - Nonprofit employment has been growing especially rapidly in the Baltimore suburbs, while growth in the city has been considerably slower between 1989 and 1996.

  - Reflecting its role as a regional capital of the nonprofit sector, the Baltimore area has disproportionate shares of large agencies, statewide or regional organizations; social service, education, health, and advocacy organizations; nonprofit health, education, and cultural employment; and agencies focusing primarily on the poor. In addition, Baltimore area nonprofits receive relatively more of their support from government, and less from either private giving or earned income.

- **Washington suburbs.** A very different nonprofit reality operates in Maryland's Washington suburbs.

  - With 36 percent of the state's population, this region accounts for about 30 percent of Maryland's nonprofit institutions and about one-fourth of the state's total nonprofit employment.

  - Though representing only 5.8 percent of total employment, nonprofit employment in the Washington suburbs grew by 38 percent between 1989 and 1996 and accounted for 37 percent of all *net job growth*.

  - Reflecting its less highly developed character, the nonprofit sector in Maryland's Washington suburbs has a larger share of young agencies; significantly smaller shares of nonprofit employment in the traditional fields of health, education, and culture; and a funding base that relies much more heavily on service fees and charges than the state as a whole. It also focuses even less heavily on those in poverty than is the case for the rest of the state.

- **Western Maryland and the Eastern Shore.** Though often associated exclusively with urban areas, the nonprofit sector is also very much in evidence in the more rural parts of Maryland in the western counties and the Eastern Shore.

  - With 15 percent of the state's population, these areas account for just over 12 percent of its nonprofit employment and around 20 percent of its nonprofit organizations. This represents half as many people as are employed in all the manufacturing industries in these regions.

  - Between 1989 and 1996, nonprofit employment grew by close to 28 percent in these two regions, compared to total employment growth of only 14 percent.

  - Compared to its counterparts elsewhere in the state, the nonprofit sector in Western Maryland and the Eastern Shore has more health providers, fewer educational institutions; and more social service, cultural, and recreational associations. In addition, it relies more heavily on government and private giving and less on earned income. What is more, it focuses less heavily on the poor than is the case for the state as a whole.

No business sector would stand for the gross lack of basic information that the nonprofit sector has had to endure in this state.

Bolstering the levels of private giving and buttressing the state's private philanthropic base are crucial to the continued vitality and independence of the state's nonprofit sector.

12. TO COPE WITH THE CHALLENGES IT FACES, THE MARYLAND NONPROFIT SECTOR IS IN NEED OF A THOROUGHGOING PROCESS OF RENEWAL INVOLVING AT LEAST FIVE CRITICAL STEPS.

- **A Maryland Civil Society Commission**--to engage opinion leaders in Maryland in a serious re-evaluation of the state's nonprofit sector and the directions in which it is headed.

- **A Monitoring System for the State's Nonprofit Sector**--to overcome the gross lack of basic information now available about the nonprofit sector in Maryland. Data developed by the Maryland Department of Labor, Licensing, and Regulation for this project provide a useful starting point for such a system. With these data made available, the state government and the state's philanthropic community should join forces to commission an annual "State of Maryland's Nonprofit Sector" report. Such a report could be featured along with existing reports on the business sector in annual roundups on the state's economy.

- **Boosting Private Giving and Buttressing the State's Private Philanthropic Base**--to overcome the limited support that Maryland nonprofit organizations receive from private charitable giving. This will require a multi-pronged approach including:

  - *Improved Tax Incentives for Giving*, such as adoption of an "above-the-line charitable deduction" for nonitemizers or the establishment of a tax credit system for generous donors.

  - *A Community Foundation Initiative* to encourage the further development of Maryland's important but still embryonic community-based philanthropic institutions.

  - *Further Fundraising Training* for existing agencies.

- *A Media Campaign to Celebrate Giving and Partnership.*

• **Building Organizational Capacity**--to further strengthen the institutional infrastructure of the nonprofit sector in the state. This will involve at least four steps:

- *Training.* Maryland still lacks an accessible, full-fledged program of nonprofit management training comparable to those that have developed in other states. A group of nonprofit leaders is now working with the Johns Hopkins Institute for Policy Studies to develop such a degree program, and these efforts deserve widespread support.

- *Technical Assistance: A Nonprofit Management Improvement Fund*--to enable nonprofit organizations to rethink and rework basic features of their organizational structure and behavior just as business organizations have recently done.

- *Improved Benefits*--to overcome the reported difficulties Maryland nonprofits are facing in maintaining adequate benefit levels for agency staff.

- *Strengthened Infrastructure Organizations*--to develop Maryland Nonprofits into an even more effective service and advocacy vehicle for the state's nonprofit sector.

• **Promoting Public Confidence.** Finally, serious steps need to be taken to retain the public's trust in the nonprofit sector. This can be done in at least two ways.

- By encouraging a more systematic process of performance measurement among nonprofit organizations; and

- By establishing standards for appropriate organizational behavior in the nonprofit sector and creating a mechanism to encourage organizations to adhere to these standards.

CONCLUSION

The State of Maryland has a vast resource for good in its nonprofit sector. The organizations that comprise this sector provide vital community services and contribute to the quality of life in dozens of other ways as well.

As this report has shown, these organizations also play an important economic role. Indeed, they have become one of the state's principal engines of job growth.

To date, however, the role of this important set of organizations has been systematically overlooked. Worse yet, that neglect has recently begun to take its toll. This is evident in significant popular disaffection, limited levels of private charitable support, significant threats to organizational effectiveness, and a slow, steady drift towards greater commercialization.

The objective of this report has been to bring this important sector out of the shadows, to document its basic scale and contours, and to identify some of the challenges it now confronts. The ultimate goal, however, is not simply to give this sector the attention it deserves, but to stimulate the actions that will allow it to achieve the promise of which it is capable. Hopefully, that task can now proceed with greater vigor.

> *Maryland still lacks an accessible, full-fledged program of nonprofit management training comparable to those that have developed in other states.*

> *The ultimate goal of this report is not simply to give Maryland's nonprofit sector the attention it deserves, but to stimulate the actions that will allow it to achieve the promise of which it is capable.*

CHAPTER ONE
INTRODUCTION

Beyond the institutions of government and private business so familiar to citizens of Maryland lies a vast collection of other organizations that play a crucial role in community life but that are largely unknown to most citizens, unreported by the media, and unexamined by either policy makers or researchers. This collection of organizations is known variously as the "nonprofit," the "voluntary," or the "charitable" sector and it includes thousands of private day care centers, adoption agencies, family counseling programs, employment and training facilities, neighborhood organizations, nursing homes, hospitals, colleges, universities, schools, self-help groups, museums, art galleries, theaters, and others.

Like business corporations, these organizations are typically privately incorporated. But unlike businesses, they do not exist principally to earn profits. Rather, they exist primarily to serve some public purpose, such as relief of suffering, the promotion of knowledge, or the encouragement of culture. Any profit they "earn" must consequently be used to support their mission and cannot be distributed to their directors or staffs.

Private, nonprofit organizations have long played a significant role in America and in Maryland. For much of the past half century, however, this role has been largely overshadowed by the dramatic growth of government. As a consequence, precious little is known in solid empirical terms about this entire set of institutions. We do not know, for example, how many nonprofit organizations exist in the state of Maryland, what these organizations do, how they finance their activities, what the level of private charitable giving is in the community, where this giving goes, what other sources of support Maryland nonprofits receive, or how these funding patterns vary among different types of agencies. Worse yet, in the absence of reliable information, this set of organizations has become enveloped in a mythology that tends to understate its true size and importance and obscure its financial structure.

Private, nonprofit organizations have long been the forgotten stepchildren of American and Maryland society.

While this situation would be a matter of concern under any circumstances, it has taken on special urgency in recent years as a consequence of the federal budget cuts that occurred in the early 1980s and again in the mid 1990s. These changes have thrust nonprofit organizations into unaccustomed prominence as the first line of defense to fill in for government cutbacks and have prompted new concerns about how to stimulate increased private charitable support. Under these circumstances, it becomes critically important to understand what the capabilities of these organizations really are, how they finance their activities, and what the realistic prospects are for a significant expansion of their role.

PURPOSE OF THIS REPORT

The purpose of this report is to fill this gap in knowledge and thus provide the factual foundation for appraising the existing role and future prospects of the nonprofit sector in Maryland. As such, it follows up on a comparable study on the nonprofit sector in the Baltimore region carried out nearly a decade ago.[1]

More particularly, the report seeks to answer the following kinds of questions:

(1) How big is the Maryland nonprofit sector? How many organizations does it contain? What are the expenditures of these organizations? How many people do they employ? What share of total state employment does this represent? How many volunteers do these organizations engage?

(2) How are these organizations distributed across the state? Are nonprofit organizations more important in some regions than others?

(3) What is the "structure" of the Maryland nonprofit sector? What types of organizations does it contain? What do these organizations do? What is their age and size structure?

(4) How do Maryland nonprofits finance their activities? What role do private giving, government support, and fees and charges play? How does this vary by field and region?

(5) How do Maryland nonprofits manage themselves? To what extent do they adhere to basic ethical standards and proper business methods?

(6) What are the major trends and issues confronting Maryland nonprofit organizations? How have these organizations coped with some of the major challenges of recent years?

FOCUS: PUBLIC BENEFIT SERVICE ORGANIZATIONS

Because the term "nonprofit sector" is rather ambiguous, it is important to be clear at the outset what the exact focus of this report is. In legal terms, a nonprofit organization is an organization, whether incorporated or not, that does not distribute profits to its directors and that serves a purpose that the U.S. Congress or state legislatures have determined to be of sufficient public interest to exempt the organization from federal or state tax liabilities.

The range of organizations that meets this legal definition is quite broad, however. It includes private professional and business organizations such as chambers of commerce and bar associations as well as day care centers, hospitals and universities. It also includes social clubs and religious congregations. Indeed, the federal tax code contains twenty six provisions under which organizations can claim tax exemption.

For the purposes of this report, we focus on the more narrow set of organizations that most people tend to have in mind when they think about the nonprofit sector--namely, those that serve essentially public purposes, i.e., that focus primarily on assisting persons other than the members of the organization. Such organizations generally secure their tax-exempt status under two of the more than 26 provisions of the U.S. tax code that provide for such exemption-- namely, sections 501(c)(3) and 501 (c)(4).[2]

To keep the project manageable, moreover, we excluded one set of public-serving nonprofit organizations--churches and other places of wor-

---

[1] Lester M. Salamon et al, *More Than Just Charity: The Baltimore Area Nonprofit Sector in a Time of Change* (Baltimore: The Johns Hopkins Institute for Policy Studies, 1990).

[2] Section 501(c)(3) grants tax exemption to organizations that are "organized and operated exclusively for religious, charitable, scientific, testing for public safety, literary, or educational purposes, or to foster national or international amateur sports competition (but only if no part of its activities involve the provision of athletic facilities or equipment), or for the prevention of cruelty to children or animals..." Section 501(c)(4) grants exemption to civic leagues or organizations not organized for profit but operated exclusively for the promotion of social welfare."

ship. Although churches are automatically classified as 501(c)(3) organizations, they perform a distinctive function that extends somewhat beyond the confines of this study, even though some of their activities resemble the social welfare activities of other nonprofit organizations. While churches, synagogues, and mosques were excluded from this study, however, we did include service organizations that are affiliated with them, such as Catholic Charities.

GEOGRAPHIC FOCUS

The geographic focus of this report is the State of Maryland. Where possible, however, an attempt was made to break the data down by region. For this purpose, five regions were defined:

- Baltimore City;
- the Baltimore metropolitan area outside the city (Baltimore, Harford, Howard, Carroll, and Anne Arundel counties);
- the Washington metropolitan area (Montgomery, Prince George's, Calvert, Charles, and St. Mary's counties);
- the Eastern Shore (Caroline, Cecil, Dorchester, Kent, Queen Anne's, Somerset, Talbot, Wicomico, and Worcester counties); and
- Western Maryland (Frederick, Allegany, Garrett, and Washington counties).

As reflected in Table 1.1, these regions are hardly equal in population. To the contrary, 48 percent of the state's population resides in the Baltimore area alone, and another 36 percent in the Washington metropolitan area. By contrast, the Eastern Shore and Western Maryland together account for only 15 percent of the state's population. Since the number and size of nonprofit organizations is likely to vary with the size of the population, we would expect that the distribution of the nonprofit sector will also be highly uneven.

METHODOLOGY

To develop the information reported here, we drew on three principal sources of data:

Maryland Department of Labor, Licensing, and Regulation (DLLR) Data. As part of its regular data gathering, the Maryland Department of Labor, Licensing, and Regulation collects quarterly reports from employers covered by Maryland's Unemployment Insurance law. Included in these reports are data on the number of employees and total wages, though no distinction is made between full and part-time employment. Most important for our purposes, the Department was able, using additional records, to differentiate between for-profit and non-profit employers and thus to

Table 1.1
Maryland Resident Population by Region, 1995

	Population	% of Total
Baltimore City	691,131	13.7%
Baltimore Metro	1,741,862	34.5%
Baltimore City & Metro	**2,432,993**	**48.3%**
DC Metro	**1,834,250**	**36.4%**
Eastern Shore	**369,182**	**7.3%**
Western Maryland	**406,013**	**8.1%**
TOTAL	5,042,438	100.0%

Source: U.S. Bureau of the Census. *USA Counties 1996.* Washington: U.S. Government Printing Office, 1996

break out employment in the nonprofit sector. What is more, these data could be broken down among major fields of activity as defined by the Standard Industrial Classification system that is used for all economic statistics. The result is a highly reliable basis for estimating the size of the nonprofit sector, at least as reflected in employment.

Survey of Nonprofit Public-Benefit Organizations. To go beyond these employment data and learn about the exact fields of nonprofit activity, the characteristics of the agencies (e.g. age, geographic service area, clientele, revenue structure, board operations), and the impact of recent budget and management developments, we conducted a mail survey of a sample of Maryland nonprofit agencies. The sampling frame for this survey was developed by combining Internal Revenue Service records on Maryland-based 501(c)(3) and 501(c)(4) organizations with statewide listings of nonprofit agencies such as the Maryland Secretary of State's Listing of Registered Charities, the Maryland Food Bank's *List of Food Pantries and Kitchens*, the Maryland Department of Human Services' *Directory of Emergency and Transitional Housing*, and the Maryland Health and Welfare Council's *Directory of Community Services in Maryland*. This procedure was used because prior research had indicated that the IRS records are seriously incomplete. In the present case, we identified approximately 14 percent more agencies in the directories than were included in the IRS data base.

Altogether, a stratified random sample of 3,228 agencies was selected for the survey. Included here were 1,000 agencies (later reduced to 987 because of duplicates) that showed up in the IRS data base as having no income or assets. Also include were 2,086 agencies that had either assets and income according to the IRS data or were identified through the directories and were not on the IRS listings. The survey was distributed in December 1995 and a follow-up completed in March of 1996.

Of these 3,228 agencies, 439 responded to the survey. This is a response rate of approximately 15 percent, lower than we hoped but sufficient to support the analysis we planned to undertake. Using the basic IRS data file, we then developed blow-up factors to adjust this sample to the overall population of agencies. Unless otherwise noted, the data reported here are the adjusted results blown up to represent the entire population of Maryland agencies. (For a more detailed discussion of the sample selection methodology, sample reliability, and blow-up factors, see Appendix A).

U.S. Census of Service Industries and Related Data. In addition to the employment data and the survey results, we also made use of the *U.S. Census of Service Industries, the Higher Education General Information Survey* (HEGIS) prepared by the U.S. Department of Education, and the American Hospital Association annual survey of hospitals to convert estimates of nonprofit employment in various industries into estimates of total expenditures. This was done by computing ratios of operating expenditures to wages in the various industries and applying these ratios to the employment and wage bill data secured for Maryland from the Department of Labor, Licensing, and Regulation. Where possible, we used Maryland ratios for these estimates. Where Maryland estimates were not available in the Census of Service Industries because of lack of coverage of an industry (e.g. hospitals) we used the most suitably available national figure on the assumption that there was no reason to expect Maryland to differ significantly from the national average on the variable of interest to us.

STRUCTURE OF THIS REPORT

The balance of this report analyzes the major findings of this research. Chapter Two, which follows, provides a broad overview of the scale of the nonprofit sector in Maryland, both overall and by region, and documents the recent trends exhibited by this crucial sector, drawing chiefly on the employment data provided through the Maryland Department of Labor, Licensing, and Regulation. Chapter Three then examines the structure of this sector in more detail, drawing chiefly on the survey results. In Chapter Four, attention turns to the revenues of the Maryland nonprofit sector and an analysis of how this varies by type, age, and size of agency. Of special concern is the relative weight of private giving, government support, and earned income in the revenues of Maryland nonprofit organizations. Chapter Five then examines the challenges facing Maryland nonprofit organizations, how agencies have responded to these challenges, and what the opinions of agency managers are about the future. In Chapter Six we zero in on one of these challenges--the extent to which nonprofit agencies adhere to basic standards of good management and ethics in their operations. Chapter Seven then brings the analysis down to the regional level focusing on the Baltimore region, the Washington region, and the balance of the state. Chapter Eight, finally, presents some general conclusions and recommendations for change.

SUMMARY

Private, nonprofit organizations have been the forgotten stepchildren of American society for some time now. Despite the neglect they have endured, however, these organizations have continued to play a vital role in community life. In Maryland as well as other states, private nonprofit organizations deliver a significant share of the health care, education, social services, counseling, advocacy, information, research, training, adoption services, nursing home care, and cultural activity that is available. With new demands being placed on these organizations and new opportunities facing them, it is time to bring them explicitly into view, to assess their contributions, to evaluate the important challenges they face, and to lay a foundation for improving their contribution to community life. It is the purpose of this report to begin such a process.

CHAPTER TWO
A MAJOR ECONOMIC FORCE

Because of the considerable growth in government spending over the past forty or fifty years, it has become commonplace to neglect the role of the private, nonprofit sector in American society and to concentrate instead on the role of the public sector. The common assumption has been that the growth of government has displaced the nonprofit sector and left it with relatively little to do. This being so, there was little reason to believe that the nonprofit sector might still constitute a significant social or economic force.

To what extent does this assumption apply to the state of Maryland? How extensive is the Maryland nonprofit, public-benefit service sector? How does this compare with other leading components of the Maryland economy, such as government and manufacturing? How does the Maryland nonprofit sector compare with its counterparts elsewhere?

The purpose of this chapter is to answer these questions. To do so, the chapter is divided into three parts. Part I examines the *overall size* of the nonprofit sector in Maryland, its relation to other components of the state's economy, and the recent trends in its development. Part II then looks at the different *components* of this sector, the role that they play in the different "subsectors" of the state's economy, and the way they have been affected by recent trends. Finally, Part III analyzes the *regional distribution* of the nonprofit sector to determine whether there are differences in the overall scale, composition, or trends affecting the nonprofit sector in different parts of the state.

The principal data source for this analysis is a unique body of data developed for this project with the aid of the Maryland Department of Labor, Licensing, and Regulation. These data provide up-to-date information on employment and wages in the Maryland nonprofit sector based on quarterly reports that Maryland employers are required to file with the state in compliance with unemployment insurance laws. Although not universal in coverage, this data source likely covers at least 90 percent of nonprofit employment.

As of 1996, Maryland nonprofit organizations employed more people than all Maryland manufacturing businesses.

What emerges most clearly from this analysis is that the nonprofit sector is a far more significant economic force in Maryland than most people realize. Even without considering its other important contributions to Maryland life, the organizations in this sector make an immense economic contribution. What is more, this contribution has been growing, and at a rate that is faster than the state's economy as a whole. This means that the nonprofit sector has, perhaps paradoxically, been one of the state's major engines of economic growth. Needless to say, this is a far different view of the Maryland nonprofit sector than the one that popular stereotypes typically convey.

Table 2.1
Paid and Volunteer Employment in Maryland's Nonprofit Sector as a Share of Total Employment, 1996

	Full-Time Equivalent Workers	
	Paid Workers	Including Volunteers[a]
Nonprofit Sector	185,088	266,612
Other sectors	2,033,105	2,075,102[b]
TOTAL	2,218,193	2,341,714
Nonprofit as share of Total	**8.4%**	**11.4%**

Sources: Employment data from Maryland Department of Labor, Licensing, and Regulation. Volunteer data based on MANO/JHU-IPS Maryland Nonprofit Survey, 1996

[a]Volunteer data based on 1994 estimates. Volunteers computed on a full time equivalent worker basis.
[b]Assumes that nonprofit volunteering represents 66 percent of all volunteering, based on Gallop surveys reported in *Nonprofit Almanac, 1992/93* (Washington: Independent Sector, 1994), p.65.

I. THE OVERALL SIZE OF THE MARYLAND NONPROFIT SECTOR

A Major Employer

The Maryland nonprofit public-benefit service sector consists of some 12,000-13,000 organizations.[1] Most of these are quite small, with few if any paid employees. Others, however, are major operations with large numbers of workers. Taken together, however, this set of organizations commands a sizable workforce of paid and volunteer labor.

One out of every 12 Maryland workers is employed by a nonprofit organization

Paid Employment. Focusing just on the paid employment, it turns out that Maryland's nonprofit organizations are one of the state's major sets of employers. In particular, as shown in Table 2.1, Maryland nonprofit organizations employed just over 185,000 workers as of the end of 1996. This represented over 8 percent of the Maryland labor force. Put somewhat differently, Maryland nonprofit organizations employ one out of every 12 paid workers in the state.

Interestingly, these estimates put Maryland almost exactly on a par with the rest of the nation in the proportionate size of its nonprof-

---

[1]This estimate is based on the number of 501(c)(3) and 501(c)(4) organizations listed in the Internal Revenue Service's Exempt Organization Master File, which stood at 12,399 as of 1994. Work on this project established, however that as many as 14-15 percent of organizations are not included on this file, either because they are too small to file the required federal Form 990, or because they are in the registration process or otherwise outside the IRS records. By comparison, the U.S. Census Bureau identified 3,899 tax-exempt organizations with at least one paid employee, and many nonprofit organizations do not meet this standard. A third listing, maintained by the Maryland Secretary of State's office, identifies some 2,389 nonprofit organizations in Maryland as of August 7, 1997. The Secretary of State's records, however, cover organizations that are raising more than $25,000 in funds from the public in Maryland each year or that retain the services of a professional solicitor or that otherwise identify themselves to that office. For U.S. Census data, see: U.S. Bureau of the Census, 1992 *Census of Service Industries*, Geographic Area Services: Maryland, SC92-A-21, (Washington: U.S. Government Printing Office, 1994), pp. 9,13.

[2]This estimate is based on an estimated Maryland population of 2,042,438 as of 1995 and a total 1996 nonprofit employment of 185,088. The comparable U.S. total numbers are 9,655,400 nonprofit employees and 263,434,000 population. U.S. nonprofit employment estimates are from: *Nonprofit Almanac: Dimensions of the Independent Sector*, 1996-97 (San Francisco: Jossey-Bass, Inc., 1997), p. 145. The U.S. population estimate is from *Statistical Abstract of the United States* (Austin: The Reference Press, 1995), p. 18.

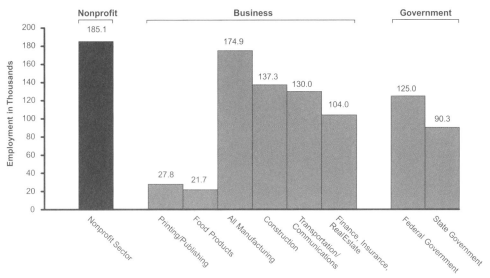

Figure 2.1 Employment in Maryland Nonprofit Sector and Selected Industries, 1996 (in thousands)

Source: Author's estimates based on data from Maryland Department of Labor, Licensing, and Regulation

it sector. Thus Maryland has 36.7 nonprofit employees per 1,000 people, almost identical to the national average.[2]

Given the structure of the Maryland economy, these figures mean that the nonprofit sector is one of the state's major "industries." Thus, as shown in Figure 2.1, employment in the nonprofit sector outdistances that in Maryland's two largest manufacturing industries (printing and food products) by a factor of more than seven times. In fact, as of 1996, the nonprofit sector employed more people than *all* Maryland manufacturing businesses put together, and it employed more people than the entire construction industry, the finance and real estate industry, the transportation and communications industry, and both the federal government and the state government. *Clearly, the nonprofit sector is a far more significant economic force in this state than is typically acknowledged.*

Volunteers. Important though it is, paid employment is not the only source of labor for Maryland nonprofit agencies. Based on the responses to our survey, Maryland nonprofit organizations also had available to them volunteer labor that is the equivalent of another 81,524 full-time employees. With volunteer effort included, the nonprofit sector thus accounts for 266,612 workers, or an even larger 11.4 percent of the state's total workforce, as Table 2.1 also shows.[3] In other words, *one out of every nine Maryland workers--paid and volunteer--works in the nonprofit sector.*

> *Maryland nonprofit organizations made expenditures of approximately $13 billion in 1996, or almost $2,600 per Maryland resident.*

---

[3]To calculate the total volunteer labor in the state, we applied estimates generated by Independent Sector and the Gallup Organization from national surveys on giving and volunteering. According to these surveys, volunteering in private, nonprofit organizations represents 66 percent of all volunteering, with the balance going to for-profit and government institutions. Total state employment was therefore increased to reflect the total volunteer time before computing the nonprofit share of the total. See: Independent Sector, *Nonprofit Almanac, 1992/3* (Washington: Independent Sector, 1992), p. 65.

Table 2.2
Wages and Operating Expenditures of Maryland Nonprofit Organizations, 1996

	Wages	Operating Expenditures
Nonprofit Sector	$5.252 billion	$12.890 billion
Total state	$66.467 billion	N.A.
Nonprofit as % of Total	7.9%	N.A.
Amount per Maryland resident	**$1,050**	**$2,578**

Sources: Employment data from special tabulations prepared by the Maryland Department of Labor, Licensing, and Regulation. Operating expenditures derived by author using ratios of wages to revenues and operating expenditures derived from *U.S. Census of Service Industries*.

A Major Source of Expenditures

This sizable employment base naturally translates into a significant impact on the state's economy. Indeed, focusing on direct salaries alone, the Maryland nonprofit sector provided over $5 billion in wages in 1996, or close to 8 percent of the state's total wages, as shown in Table 2.2. This represents more than $1,000 per Maryland resident.

But wages are only one measure of the economic impact of this sector. Nonprofit organizations made other expenditures as well. In fact, we estimate the overall operating expenditures of Maryland nonprofit organizations in 1996 to be close to $13 billion. This translates into the equivalent of almost $2,600 per Maryland resident.

Recent Trends--A Dynamic Sector

Not only is the nonprofit sector a major economic presence in the Maryland economy, but also it is a growing presence. In fact, as Table 2.3 shows, *the growth of nonprofit employment has actually outpaced the growth of state employment as a whole.*

Table 2.3
Growth of Maryland Nonprofit Employment, 1989-96

	Employees 1989	Employees 1996	Change, 1989-96 Amount	Change, 1989-96 %
Nonprofit Organizations	149,785	185,088	+35,303	+24
All Industries	2,150,918	2,218,193	+67,274	+3
Nonprofit as % of Total	7.0%	8.3%	52%	

Source: Author's computations based on special tabulations provided by the Maryland Department of Labor, Licensing and Regulation

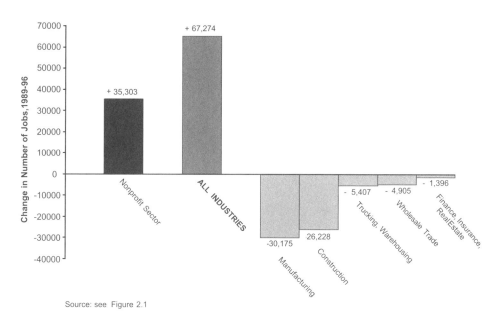

Figure 2.2 Changes in the Number of Jobs in the Nonprofit Sector and Other Industries in Maryland, 1989-96

Source: see Figure 2.1

- Between 1989 and 1996, the nonprofit sector added more than 35,000 new employees in Maryland, an increase of 24 percent. This translates into an annual average growth rate of 3.0 percent per year.

- By contrast, during this same period, the entire state economy added only 67,274 new jobs, an increase of only 3 percent. This represents an annual average growth rate of only 0.4 percent per year. *The nonprofit sector thus accounted for over half of the state's total net job growth over this seven-year period.*

- This growth in nonprofit employment is all the more striking in comparison to trends in many other state industries. Thus, compared to the 35,303 net jobs added by the nonprofit sector between 1989 and 1996, Maryland's manufacturing industry lost 30,175 jobs, its construction industry lost 26,228 jobs, its trucking and warehousing industry lost 5,407 jobs, its wholesale trade industry lost 4,905 jobs, and its finance, insurance, and real estate industry lost 1,396 jobs (See Figure 2.2).

Nonprofit organizations accounted for over half of the total net job growth in Maryland between 1989 and 1996.

While outpacing the growth of employment generally in Maryland between 1989 and 1996, nonprofit employment grew more robustly in the early part of this period than during the more recent part.

- Thus, as shown in Figure 2.3, nonprofit employment grew at an annual average rate of 4.9 percent per year during the period 1989-92, but then declined to 1.8 percent per year during 1992-95 and to only 1.0 percent between 1995 and 1996.

- This pattern is all the more striking in comparison to that for total state employment,

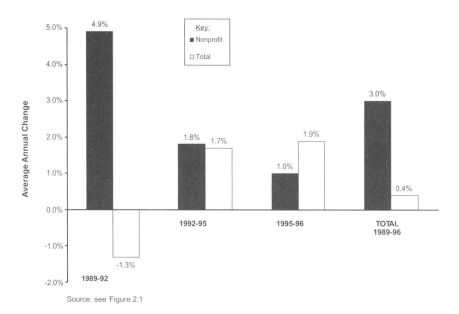

Figure 2.3 Average Annual Growth Rates of Nonprofit Employment and Total Maryland Employment, 1989-92, 1992-95, 1995-96

Source: see Figure 2.1

which declined during the early period and has forged ahead of nonprofit employment growth in the past year (1995-96).

Several conclusions seem to flow from these data:

- In the first place, it appears that the nonprofit sector may serve at least in part as a countercyclical mechanism in the Maryland economy, increasing in scale as economic recession reduces for-profit employment.

- In the second place, however, the sector's ability to continue contributing to state economic growth seems to have hit a ceiling.

- One reason for this may be the reductions in government support that occurred in the mid-1990s. As we will see, such income is extremely important to the fiscal health of this sector.

- Equally important were very likely the immense shifts that occurred in the medical field during this period, which have put increased pressures on nonprofit health providers. Since health providers comprise a significant share of total nonprofit employment, these developments may have had a particularly chilling effect.

- Finally, for-profit firms have increasingly entered markets that nonprofits have traditionally dominated, exposing nonprofit providers to significant competitive pressures. These shifts may be affecting the growth rates of nonprofit organizations.

To evaluate these various factors, however, it is necessary to turn from this general overview of the nonprofit role in the Maryland economy to a more specific analysis of the *components* of the nonprofit sector and at the trends within the separate *subsectors* that comprise it.

II. TYPES OF NONPROFIT ORGANIZATIONS

While the organizations that comprise the nonprofit sector share a number of crucial features, they are hardly identical. To the contrary, this sector is composed of a variety of subsectors that differ significantly in size, character, and dynamics. Overall aggregates can therefore obscure important differences among the subsectors. To understand the real character of the state's nonprofit sector, therefore, it is necessary to go beyond the aggregate figures and look at the various subsectors in more detail.

Over half the people employed in the nonprofit sector in Maryland work in the health field.

Composition of the Maryland Nonprofit Sector

As a first step in this direction, Figure 2.4 records the breakdown of total nonprofit employment among the major subcomponents of the Maryland nonprofit sector. As this figure makes clear, nonprofit employment in Maryland is hardly divided evenly among the subsectors. In particular:

- **Health.** The health subsector clearly dominates the nonprofit scene in Maryland in terms of employment. Over 50 percent of all people working in this sector are employed by health organizations, most of them (42 percent) in hospitals. This reflects a long tradition of provision of health services through private, nonprofit organizations, both religious and nonreligious.

- **Education and research.** The second largest component of the Maryland nonprofit sector is education and research, which accounted in 1996 for 21 percent of nonprofit employment in the state. Included here is higher education (8.6 percent), research (4.5 percent) and elementary and secondary education (7.8 percent).

- **Social services.** The third largest component of the nonprofit sector in terms of employment is social services, with 19 percent of total nonprofit employment. Included here

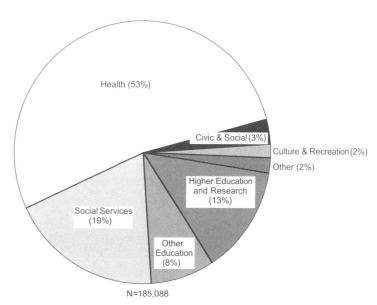

Figure 2.4 Employment in Maryland Nonprofit Organizations, by Field, 1996

Source: see Figure 2.1

is residential care (8.3 percent), individual and family services (3.6 percent), child day care (2.5 percent), job training (2.2 percent), and other social services (2.6 percent).

- **Culture and Arts and Civic.** At the opposite extreme, culture and arts organizations and civic and social organizations account for much smaller shares of total nonprofit employment (about 2 percent each).

This general structure of the nonprofit "industry" in Maryland is quite close to the picture nationally. Thus, as shown in Table 2.4 below, there are only marginal differences between Maryland and the nation in terms of the shares that the different subsectors comprise of the total nonprofit sector. Thus, for example, compared to the 53 percent of all nonprofit employment in the health subsector in Maryland, the national figure is 51 percent. Similarly, compared to the 21 percent of all nonprofit employment in education and research in Maryland, the national figure is 23 percent. Although some deviations are apparent in the data (e.g. Maryland seems to have proportionately more people employed by nonprofit research institutions and fewer in higher education institutions than is the case nationally), the overwhelming story is one of striking consistency between the composition of the Maryland nonprofit sector and the composition of this sector at the national level.

Nonprofit Presence in Key Service Fields

Given this breakdown of nonprofit employment among industries, it should come as no surprise that the nonprofit sector is far more important in some "industries" or fields than its share of overall state employment might suggest. Thus, as Figure 2.5 shows, compared to its 8 percent share of total state employment, nonprofit organizations account for:

- 9 percent of *culture and recreational* workers;

- 13 percent of *elementary and secondary education* employment;

- 34 percent of *higher education* employment;

- 48 percent of employment in *health*; and

- 54 percent of employment in the field of *social services*.

Nonprofit Trends by Field

Not only do the various components of the nonprofit sector differ in size, but also they differ in their recent rates of growth. Thus, while

Table 2.4
Nonprofit Sector Employment, by Industry, Maryland vs. U.S. Average

Subsector	Percent of Total	
	Maryland	U.S.
Health	53%	51%
Education, research	21	23
Social and legal services	19	18
Civic, social, fraternal	5	5
Culture and recreation	2	2
Total	100%	100%

Source: U.S. figures from Bureau of Labor Statistics *Employment and Earnings*, 1996; and *U.S. Census of Service Industries* as reported in *Nonprofit Almanac, 1996-97*. Maryland data based on statistics made available from the Maryland Department of Labor, Licensing, and Regulation.

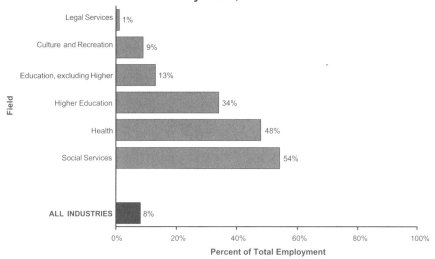

Figure 2.5 Nonprofit Share of Maryland Employment, by Field, 1996

Sources: Author's estimates based on data from Maryland Department of Labor, Licensing, and Regulation; US Census of Service Industries, 1992; US Census of Government, 1995

the nonprofit sector as a whole outpaced the growth of the overall Maryland economy between 1989 and 1996 by a substantial margin (24 percent versus 3 percent), some components of the nonprofit sector did even better than the average while others lagged behind. Interestingly, however, in all but one field the rate of growth of nonprofit employment exceeded that for state employment as a whole. In particular, as reflected in Figure 2.6 below:

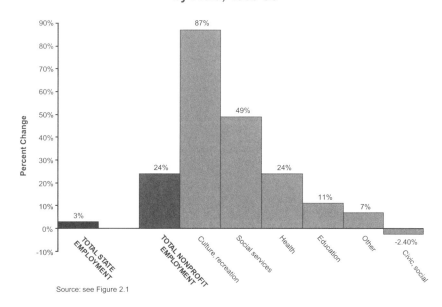

Figure 2.6 Changes in Nonprofit Employment in Maryland by Field, 1989-96

Source: see Figure 2.1

- **Arts and recreation.** Nonprofit culture, arts, and recreation employment grew by an astounding 87 percent over this seven-year period, or at an annual average rate of 9 percent per year. This was nearly four times the growth rate for nonprofit employment as a whole, and nearly 30 times greater than the growth rate of overall state employment. Most of this growth took place during the 1989-92 period, when arts employment increased by close to 14 percent per year. Arts and recreation employment grew at a robust rate of 5.2 percent a year during the subsequent 1992-96 period as well, however.

- **Social services.** Nonprofit social service agency employment also grew at a rate well above the sector average during this period. Nonprofit social service agency employment grew by almost 50 percent during this period, a growth rate of 5.7 percent per year. Here, again, growth was particularly strong during the 1989-92 period, as federal social service spending resumed its rise after a period of decline and stagnation. Nonprofit social service agency employment grew at an annual average rate of 8.4 percent during this period. Such growth continued at a rate of 4.8 percent per year during the 1992 to 1995 period, moreover, before plummeting to under 1 percent in the aftermath of the 1994 Congressional elections and the resumption of a period of budget cuts.

- **Health services.** The growth rate of health employment was more modest during this period. Such employment grew by 24 percent between 1989 and 1996, or at a rate of 3 percent per year. As in several other fields, moreover, the growth rate in the health field was much stronger during the early part of this period (1989-92) than during the latter part (1992-96), when it dropped off to just under 2 percent per year. What is more, a significant disparity was evident between the growth of hospital employment and the growth of other health services. The latter grew by an impressive 114 percent between 1989 and 1996, while the former grew by only 11 percent. Indeed, between 1992 and 1996, nonprofit hospital employment actually declined in the state.

Virtually every component of the nonprofit sector added jobs at a rate that exceeded the growth rate of overall employment in the state between 1989 and 1996.

- **Education.** The education component of the nonprofit sector registered even less growth during this period. Overall, nonprofit education employment grew by only 11 percent between 1989 and 1996, or at an average annual rate that was only half as great as that of the sector as a whole (1.4 percent vs. 3.0 percent). Even this overstates the recent trends, however, since the period of 1992-95 witnessed an absolute decline in nonprofit educational employment. Much of this decline was concentrated, however, in higher education. Indeed, higher education employment declined overall by 2 percent between 1989 and 1996, while education employment outside of higher education grew quite robustly (up 29 percent).

One out of every five jobs in Baltimore City is in the nonprofit sector.

- **Civic, Social, and Other.** Civic and social organizations and other organizations did even worse during this period, the former actually losing employment on average and the latter growing at an annual average rate of barely 1 percent.

Given these differential growth rates, the contributions that various components of the nonprofit sector made to the sector's overall growth during this period differed significantly, both from each other and from their respective shares of overall sector size at the start of the period. Thus, as Figure 2.7 shows:

- Social service organizations accounted for 33 percent of the growth between 1989 and 1996 even though they accounted collectively for only 18 percent of total nonprofit employment when the period started in 1989.

- Culture and recreation organizations accounted for 4 percent of the growth even though they represented just over 1 percent of total nonprofit employment when the period started.

- By contrast, education, which accounted for 20 percent of nonprofit employment as of 1989, contributed only 8 percent of the growth.

- Even the health component failed to grow at a rate greater than its share of total employment. Rather, its share of the growth was exactly the same as the share of the total it held in 1989, i.e. 53 percent.

What is clear from these data is that the Maryland nonprofit sector is becoming somewhat more diverse. At least in terms of employment, culture, arts, recreation and social services are gaining ground while education, civic, and health organizations are falling back. While health may still dominate the Maryland nonprofit scene, other components seem to be narrowing the gap.

In the larger fields of nonprofit action--i.e. health, education, and social services--nonprofit organizations appear to be losing market share despite often substantial growth.

The Changing Position of the Nonprofit Sector vis-a-vis For-Profit Providers

Even this picture is somewhat incomplete, however, since the nonprofit sector does not operate in isolation. Rather, it is in an increasingly competitive position vis-a-vis for-profit firms in many of the tra-

Figure 2.7 Share of Nonprofit Employment Growth in Maryland, 1989-96, by Field

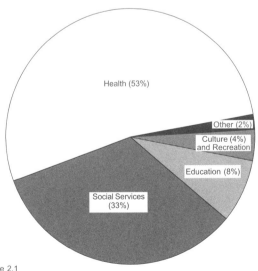

Source: see Figure 2.1

ditional fields of nonprofit activity. How are nonprofit providers faring in this competition? Are the growing sectors growing only in absolute terms while losing ground relative to for-profit providers in the same field? Or are they managing to fend off the competition?

Table 2.5 provides a first approximation of the answer to this question. This table compares the proportion of private employment in each field accounted for by the nonprofit sector in 1996 to the proportion it accounted for in 1989. Where this ratio is greater than 1.0 it indicates that the nonprofit sector gained ground on the for-profit sector in a field. Where the ratio is less than 1.0 it indicates that nonprofit organizations lost ground relative to for-profits.

As this tables shows:

- Overall, nonprofit organizations gained ground on the private for-profit sector during this period. Thus, the nonprofit share of total private employment was proportionately 20 percent higher in 1996 than it had been in 1989.

- Much of this performance was concentrated in three fields, however, where for-profit competition is relatively limited--arts and recreation, civic and social organizations, and the broad "other" category.

- In the larger fields of nonprofit action--i.e. health, education, and social services--nonprofit organizations appear to be losing market share despite often substantial growth. This is particularly striking in the field of social services, where nonprofit organizations suffered a 6 percent decline in their market position relative to for-profit companies.

- The situation is somewhat more complicated in the education field than these data suggest, moreover, because the chief competition for nonprofits in this field is from governmental institutions rather than for-profit ones. And from the evidence at hand, that competition has been significant. Thus, while representing 20 percent of total educational employment as of 1989, nonprofits accounted for only 14 percent of educational employment growth between 1989 and 1996.

While these data show an increasingly competitive environment for nonprofit activity, however, they also suggest that nonprofit providers in Maryland appear to be holding their own better than their counterparts elsewhere. Thus, while the nonprofit share of health employment has declined at the national level, it has remained relatively steady in Maryland. Maryland social service providers, however, are experiencing pressures very similar to those elsewhere. Between 1977 and 1992, for example, for-profit providers accounted for 31 percent of total private social service

Table 2.5
Change in Nonprofit Share of Private Employment, by Field, 1996 vs. 1989

Subsector	Ratio of Nonprofit Share of Private Employment in 1996 to Share in 1989
Culture and recreation	1.47
Other	1.15
Civic, social	1.04
Health	.99
Education	.99
Social services	.94
Total	**1.21**

Source: See Table 2.3

employment growth nationally though they started the period with only 21 percent of the employment. In Maryland over the more recent 1989-96 period, for-profits accounted for 40 percent of the private employment growth though starting the period with only 25 percent of the total. Clearly, Maryland nonprofits in at least some fields are beginning to experience serious competition from for-profit providers in their traditional areas of activity.

III. REGIONAL DISTRIBUTION

Not only does the nonprofit sector vary considerably from subsector to subsector, moreover, it also varies considerably from area to area. It is quite possible, therefore, that the aggregate trends identified above may conceal important variations by region. More than that, the patterns of nonprofit growth or decline identified above may really result from broader economic or demographic trends affecting particular regions of the state. To understand what is happening to this sector, therefore, it is necessary to go beyond the aggregate and compositional data examined so far and analyze the patterns by region.

For this purpose it is useful to divide the state of Maryland into five regions: Baltimore City, the Baltimore suburbs outside the city, the Washington suburbs, the Eastern Shore, and Western Maryland.[4]

The Aggregate Picture

Distribution of Nonprofit Employment. As Figure 2.8 indicates, the nonprofit sector is present in all portions of the state, but, like the state's population more generally, its "capital" is clearly Baltimore City and the Baltimore region more generally. In particular:

- 39 percent of all Maryland nonprofit employees work in Baltimore City. This is almost three times larger than Baltimore City's share of the state's total population. In addition, 25 percent of the state's nonprofit

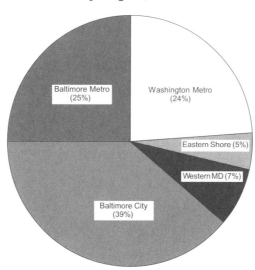

Figure 2.8 Distribution of Maryland Nonprofit Employment, by Region, 1996

N=185,088
Source: see Figure 2.1

[4]For the purposes of this analysis, the Baltimore suburbs include Baltimore County, Anne Arundel County, Howard County, Carroll County and Harford County; the Washington suburbs include Prince George's, Montgomery, St. Mary's, Calvert, and Charles; the Eastern Shore includes Caroline, Cecil, Dorchester, Kent, Queen Anne's, Somerset, Talbot, Wicomico, and Worcester counties; and Western Maryland includes Frederick, Allegany, Garrett, and Washington counties.

employees work in the Baltimore metropolitan area outside of the city. Thus, with 48 percent Maryland's total population, the Baltimore area accounts for 64 percent of the state's nonprofit employees.

- Another significant concentration of nonprofit employment in Maryland is in the Washington suburbs. With 36 percent of the state's population, this area accounts for 24 percent of the state's nonprofit employment.

Nonprofit Share of Total Employment. Reflecting this pattern of distribution, nonprofit employment plays a more significant role in the labor force of Baltimore City than in other regions of the state. At the same time, this sector is a significant presence in every region. In particular, as shown in Figure 2.9 below:

Even in Western Maryland and the Eastern Shore, the nonprofit sector represents a significant 7-8 percent share of total employment--employing half as many people as manufacturing and more than construction, transportation, wholesale trade, and finance and insurance.

- Nonprofit organizations account for 18 percent of total employment in Baltimore City. In other words, one out of every five jobs in Baltimore City is in the nonprofit sector.

- Elsewhere the nonprofit share of total employment is considerably lower than this. However, even in Western Maryland and the Eastern Shore, the nonprofit sector represents a significant 7-8 percent share of total employment. Put somewhat differently, the nonprofit sector employs over half as many workers in these two regions as manufacturing, and more workers than most of the other industries in these regions, including construction, transportation, wholesale trade, and finance, real estate and insurance.

- Although nonprofit employment is only 6 percent of total employment in the

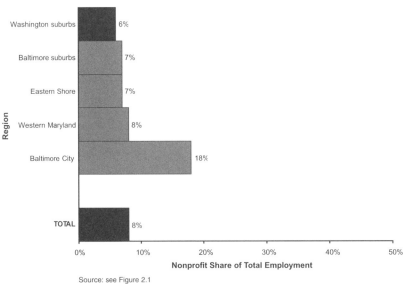

Figure 2.9 Nonprofit Share of Total Employment in Maryland, by Region, 1996

Source: see Figure 2.1

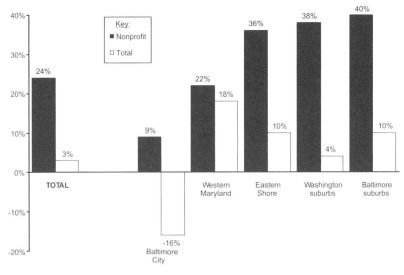

Figure 2.10 Growth of Nonprofit and Total Employment in Maryland, by Region, 1989-96

Source: see Figure 2.1

Washington suburbs, moroever, it still outdistances manufacturing, transportation, wholesale trade, and finance and insurance.

Recent Trends. Not only is nonprofit employment an important part of the economies of the state's different regions, it has been an even more important part of the regions' economic growth. In fact, in every region the growth of nonprofit employment has outpaced the growth of total employment. More specifically, as Figure 2.10 shows:

In every region the growth of nonprofit employment has outpaced the growth of total employment.

- Nonprofit growth was slowest in Baltimore City. Even here, however, it grew by 9 percent--three times faster than overall state employment growth during this period. By comparison, total employment in Baltimore City declined by 16 percent. The nonprofit sector thus accounted for all of the net growth that occurred.

- Elsewhere, nonprofit employment growth was even greater--22 percent in Western Maryland, 36 percent on the Eastern Shore, 38 percent in the Washington suburbs, and 40 percent in the Baltimore suburbs. By contrast, the highest rate of overall employment growth was only 18 percent (in Western Maryland).

- In no region, therefore, did the overall rate of employment growth equal or exceed the growth rate of nonprofit employment. In other words, the nonprofit sector has been one of the engines of overall employment growth in every region. Even in Baltimore City, where overall employment declined, the nonprofit sector grew, albeit at a rate that lagged behind the other regions. In the more rapidly growing other regions as well, however, nonprofit employment outpaced the growth rate of

Table 2.6
Shares of Nonprofit Employment Growth in Maryland, by Region, 1989-1996

	(1) Share of Nonprofit Employment, 1989	(2) Share of Nonprofit Employment Growth, 1989-96	(3) Ratio of Share of Growth to Total Share in 1989 (2/1)
Baltimore City	43.9%	16.1%	0.37
Western Maryland	7.2	6.7	0.93
Eastern Shore	4.7	6.9	1.47
Washington suburbs	21.5	33.3	1.55
Baltimore suburbs	22.7	37.0	1.63
Total	100.0%	100.0%	

Source: See Table 2.3

total employment, so that nonprofit employment increased as a share of the total.

What these data make clear is that a significant decentralization appears to be underway within the state's nonprofit sector, as nonprofit organizations follow the state's population to the outlying areas. Table 2.6 above demonstrates this point even more vividly. As this table shows:

- Baltimore City, which contained 44 percent of the state's nonprofit employment as of 1989 captured only 16 percent of its growth between 1989 and 1996. Its growth was thus only 37 percent of what its starting share would have suggested.

- By contrast, the suburbs of Baltimore, which claimed only 23 percent of state nonprofit employment as of 1989, captured 37 percent of the growth, 1.6 times more than its proportional share.

In every region the growth of nonprofit employment has outpaced the growth of total employment.

- The Washington suburbs and the Eastern Shore also increased their nonprofit employment at rates that were disproportionately high given their shares of the total when the period began.

As the state's residents have spread out to the suburbs and the Shore, they have created a demand not only for for-profit businesses, but also for nonprofit institutions. Baltimore City's dominance of the state's nonprofit sector is thus declining as competing centers of nonprofit action emerge in other regions as well.

The implications of this development are obviously great. For one thing, it means that the the nonprofit sector is clearly becoming a statewide "industry" rather than a more nearly regional one. This has obvious consequences for the political clout as well as the visibility of this set of institutions. At the same time, this development indicates that the traditional role of the urban core as the center of culture and community life is waning. This means that traditional urban cultural and social welfare institutions can expect increased competition from suburban counterparts for the available private, charitable dollars. In addition, millions of dollars of resources are being absorbed reproducing in the suburbs the cultural, social welfare, and related institutions and infrastructure that already exist in the city. Whether this is a sensible use of limited resources is a question that all the state's residents must ponder.

Variations by Subsector

Before we can be sure about the full meaning of these trends, however, it is necessary to disaggregate them a bit to determine whether variations exist in the composition of the nonprofit sector in different regions. Conceivably, some specialization may be occurring within the nonprofit sector, with different regions concentrating on particular types of services or activities. To the extent this occurs, certain efficiencies may be created with regard to the operation of the state's nonprofit institutions.

To evaluate this, we computed "concentration ratios" for each segment of the nonprofit sector for each region. The concentration ratio is the total ratio of the share of nonprofit employment in a particular field in a region to the share that this field represents in the state. Thus, for example, if a particular region has 40 percent of its nonprofit employment in social services whereas the state average is 20 percent, the concentration ratio for social services for that region would be 40/20, or 2.00.

As Table 2.7 below indicates, significant differences do exist in the composition of the nonprofit sector from region to region. In particular:

Baltimore City functions as a regional center for cultural, educational, and health activities for the metropolitan region and, to a certain extent, for the state as a whole.

- Baltimore City has a far higher concentration of nonprofit employment in health, education, and culture and recreation than do the other regions or the state as a whole. This likely reflects the presence in Baltimore City of several private universities, hospitals, schools, and cultural institutions, including particularly Johns Hopkins University and Medical Center, perhaps the state's large private employer, the symphony, several museums, and other cultural institutions. By contrast, the city has far lower concentrations of nonprofit employment in social services, civic and social activity, and other functions.

The nonprofit sector is not a "single" thing in the state of Maryland. Its role and function differ from place to place.

What this suggests is that the city is functioning as a regional center for cultural, educational, and health

Table 2.7
Composition of the Nonprofit Sector, by Industry, by Region, 1996

	Concentration Score[a]					
Region	Health	Educa.	Social Svcs.	Rec. Culture	Civic Social	Other
Baltimore City	1.13	1.40	.57	1.77	0.37	0.35
Baltimore suburbs	0.85	0.85	1.28	0.44	1.28	1.76
Washington suburbs	0.89	0.71	1.32	0.54	1.58	1.47
Eastern Shore	1.07	0.69	1.23	0.53	1.22	0.53
Western Maryland	1.14	0.61	1.06	0.31	1.31	0.67
State Average	**52.8%**	**16.4%**	**19.5%**	**1.6%**	**3.4%**	**6.3%**

Source: See Table 2.3
[a]Concentration Score = Ratio of percent of total nonprofit employment in a field in the region to the share of total nonprofit employment in the field in the State, 1996.

activities for the metropolitan region and, to a certain extent, for the state as a whole. Even in areas such as social services where its share of total nonprofit employment lags behind the state's as a whole, moreover, Baltimore City is still probably functioning as a regional center since the number of social service agency employees per 1,000 residents is 60 percent higher in Baltimore City than in the state as a whole (11.4 vs. 7.1).

- In the suburban areas of Baltimore and Washington, the nonprofit sector seems to take a somewhat different character from what it takes in Baltimore City:

  - While health services are still dominant, they do not dominate to the extent they do in the city. Thus the health share of total nonprofit employment is only roughly 85-90 percent as great in the suburban areas as in the state as a whole. Certainly for the more serious health treatments, the large urban-based nonprofit hospitals continue to dominate the metropolitan area.

  - Similarly, nonprofit education, culture and recreation services are considerably less pronounced in the suburbs than in the city.

  - By contrast, the nonprofit sector is proportionally far more in evidence in these regions in the fields of social services and civic and social life. The former is somewhat surprising since social services are commonly associated with the problems of the poor, and the poor are clearly more heavily concentrated in the urban core. What the data here suggest, however, is that social services are now an increasingly middle class service field, with nursing home care, family counselling, drug abuse prevention, day care, and related activities increasingly appealing to a middle class clientele in the suburbs. What this phenomenon will mean for traditional poverty-focused social service provision is anyone's guess, but it is clear from the data reported here that we are witnessing an immense suburbanization of social service provision, and hence of social service agencies.

- The profile of the nonprofit sector differs again in the more outlying areas of Western Maryland and the Eastern Shore. As in the suburban jurisdictions, the nonprofit sector in these regions is more heavily concentrated in civic and social life and social services than either culture and recreation or education. Unlike these jurisdictions, however, the health share of nonprofit employment is at or above the state average, presumably reflecting the inaccessibility of urban-based health institutions in these regions and the need to establish local institutions instead.

Baltimore City is losing its role as the regional center of nonprofit activity not only in the population-sensitive areas of civic and social organization but also in health, education, and culture.

In short, the nonprofit sector is not a "single" thing in the state of Maryland. Its role and function differ from place to place. Baltimore City functions as the home of a number of large-scale regional nonprofit institutions providing health, cultural, and educational services to a regional and statewide population. Elsewhere, the sector plays a more limited service and social role for local populations.

Recent Trends. This general pattern of nonprofit specialization by region is hardly stable, however. Not only do the regions vary in the shape of their nonprofit sectors, but these patterns are highly dynamic.

To illustrate this, we computed a "relative growth score" for each field of nonprofit action

Table 2.8
Relative Share of Nonprofit Growth, 1989-96, by Field, by Region

Region	Relative Growth Score[a]						
	Health	Education	Social Svcs.	Rec. Culture	Civic, Social	Other	Total
Baltimore City	0.55	0	0.68	0.78	0	0	0.37
Baltimore suburbs	1.36	2.22	1.82	1.08	2.76	0.30	1.63
Washington suburbs	1.83	3.40	0.74	4.50	0	2.68	1.55
Eastern Shore	0.89	4.31	1.05	0.87	3.50	122.22	1.47
Western Maryland	0.91	0.37	0.51	0.83	3.39	2.11	0.93
TOTAL	0.96	0.49	1.78	3.70	0.58	0.75	-

Source: See Table 2.3

[a] Relative Growth Score = ratio of region's share of nonprofit employment growth between 1989 and 1996 to region's share of total nonprofit employment in field as of 1989.

in each region. The relative growth score compares the share of total nonprofit employment growth in a field that a particular region accounted for during 1989-96 to the share of total nonprofit employment in the field that the region represented as of the beginning of the period. Thus, for example, if a region accounted for 20 percent of the growth in nonprofit health employment in the state during 1989-96 but represented 40 percent of all nonprofit employment in health as of 1989, it would have a "relative growth score" of 20/40= 0.50. Generally speaking, a growth score less than 1.00 indicates that a region is not keeping pace with its prior level of employment in a field of nonprofit action and a growth score greater than 1.00 indicates that it is gaining ground relative to other regions as a provider of the service.

Table 2.8 records the results of this analysis. What it shows is the following:

- In the first place, it is clear from this table that the relatively lower rate of growth of nonprofit employment in Baltimore City compared to the state as a whole is not the product simply of a single field. Rather, Baltimore City's relative growth score is less than 1.00 in every field of nonprofit action. Baltimore City is thus losing its role as the regional center of nonprofit activity not only in the population-sensitive areas of civic and social organization but also in health, education, and culture.

While Baltimore City is slipping as a regional center of nonprofit action, the suburban areas of Baltimore and Washington are rapidly gaining ground.

- In three fields (education, civic and social, and other), Baltimore City not only did not contribute proportionally to overall state nonprofit employment growth but actually lost employment.

- Even in the field of recreation and culture, where nonprofit employment grew by 61 percent in Baltimore City between 1989 and 1996, this was still below the 79 percent share of such employment that Baltimore City claimed as of 1989, when the period began.

- In the health field, Baltimore City claimed only half of the share of growth that its proportion of the total would have suggested.

- While Baltimore City is slipping as a regional center of nonprofit action, however, the suburban areas of Baltimore and Washington are rapidly gaining ground.

  - In the case of the Baltimore suburbs, this was true in almost every field of nonprofit action--health, education, social services, recreation and culture, and civic and social action.

  - In the case of the Washington suburbs, it was true in every field but two--social services and civic--with growth particularly marked in education and culture and recreation.

  - To some extent, these shifts represent not the creation of wholly new nonprofit institutions but the suburbanization of existing city-based organizations. This is the case, for example, with the dramatic extension of the Johns Hopkins Medical System into the Baltimore suburbs, the creation of a Johns Hopkins University teaching facility in Montgomery County, and the extensions of several Baltimore City-based social service agencies into the surburbs.

- Though not as fully developed as in the suburban areas, a similar spreading of nonprofit action is also evident on the Eastern Shore, especially in the fields of education, social services, and civic and social life.

Like other facets of society, therefore, the Maryland nonprofit sector is "suburbanizing," following the movement of people out of central cities to the suburbs. This pattern is consistent with national trends, but raises significant questions about the new infrastructure that nonprofit institutions must build in their new locations and about the help that will be available for the people left behind.[5] At the same time, these developments suggest that recent fears about the decline of "social capital" in America may be somewhat overstated.[6] In the State of Maryland, at least, there is ample evidence that residents are re-creating in the suburbs the infrastructure of nonprofit institutions, especially in the civic and social area, that they left behind in the cities.

CONCLUSION

The nonprofit sector is thus a sizable and dynamic presence in the Maryland economy. Without even considering the contribution of volunteers and the numerous non-service functions that this sector performs, it is clear that nonprofit organizations make a major contribution to the economy of this state, providing a crucial range of human services and employing sizable numbers of people. Although this sector is particularly evident in the Baltimore area, moreover, its presence is felt elsewhere as well--in the Washington suburbs, in the western counties, and on the Eastern Shore. In fact, the sector's growth has been particularly striking in the suburban areas as it has followed populations out of the urban core.

Important as this assessment of the economic position of the nonprofit sector is, however, it only captures a portion of this sector's character and role. Of the 12,399 valid 501(c)(3) and 501(c)(4) organizations listed for Maryland in the Internal Revenue Service's data base of tax-exempt entities, for example, close to 7,500 reported no income or assets as of 1994. Clearly, it is necessary to go beyond measures of employment and expenditures to capture the full contribution and nature of this set of institutions. It was for this reason that we supplemented these employment measures with a survey of the state's nonprofit organizations. It is to the results of this survey, and a more in-depth picture of the character of the state's nonprofit organizations, that we therefore now turn.

---

[5]For a discussion of the suburbanization of the nonprofit sector nationally, see Julian Wolpert, *Patterns of Generosity in America: Who's Holding the Safety Net?* (New York: The Twentieth Century Fund Press, 1973), pp. 28-29.

[6]Robert Putnam, "Bowling Alone: America's Declining Social Capital," *Journal of Democracy,* Vol. 6, (1995), pp. 65-78.

CHAPTER THREE
A CLOSER LOOK: THE STRUCTURE AND FOCUS OF MARYLAND'S NONPROFIT SECTOR

As the previous chapter made clear, Maryland's nonprofit sector is an important economic force in the state, accounting for one out of every 12 paid workers, one out of every 9 paid and volunteer staff, and half of the state's recent net employment growth. But what do the nearly 13,000 organizations that comprise this sector really do? How is the sector structured? How old are these agencies? How are they distributed in terms of size? Whom do they serve? And what really lies behind the sector's recent growth?

Although Chapter Two above was able to shed some useful light on the answers to these questions with the help of newly available state employment data, these data leave important gaps in our knowledge. The purpose of this chapter is to fill in some of these gaps. To do so, it draws on a survey of Maryland nonprofit organizations carried out in 1996 by the Maryland Association of Nonprofit Organizations in association with this author and the Johns Hopkins Institute for Policy Studies.

As noted in Chapter One above, this survey was distributed to a stratified random sample of nonprofit, public-benefit organizations other than religious congregations in Maryland. The sample frame was constructed by combining listings of nonprofit organizations available from the U.S. Internal Revenue Service, the Maryland Secretary of State's office, and statewide information and referral service manuals listing nonprofit organizations in a variety of fields. To ensure a sufficient representation of the larger organizations, which are fewer in number, such organizations were oversampled. Altogether, survey forms were distributed in December 1995 to approximately 3,000 agencies, of which 439 ultimately responded. These responses were then blown up to represent the universe of nonprofit agencies in the state. (See Appendix A for a detailed discussion of the methodology).

Perhaps the most salient characteristic of Maryland's nonprofit sector is its diversity...There are few areas of community life in which nonprofit organizations are not making a contribution in Maryland.

In the balance of this chapter we examine some of the major features of the Maryland nonprofit sector that emerge from these survey responses. We begin by looking at the activities in which these organizations are engaged and then examine the age and size structure of the sector. Finally, we examine the sector's client base. Against this backdrop, Chapter Four then examines the revenue structure of Maryland nonprofit organizations, which has importantly shaped the sector's structure and role.

I. WHAT MARYLAND NONPROFIT ORGANIZATIONS DO

Basic Activities

Perhaps the most salient characteristic of Maryland's nonprofit sector that emerges from our survey responses is its diversity. While the employment data reported in Chapter Two groups agencies under a handful of Standard Industrial Classification codes, it is clear from the survey responses that this obscures a tremendous amount of detail about what Maryland nonprofit organizations actually do. In practice, these organizations contribute to the quality of Maryland life in far more numerous ways:

- over 2,000 provide some kind of emergency food and clothing
- over 700 offer child day care services
- more than 500 offer housing assistance
- more than 2,000 make some contribution to environmental protection
- close to 1,000 are involved in ethnic or group awareness.

Indeed, as Appendix B shows in great detail, there are few areas of community life in which nonprofit organizations are not making a contribution in Maryland. In fact, most agencies are involved in multiple fields. Indeed, over half of the agencies reported some activity in *three* or more of the 14 major fields we identified in our survey. Maryland's nonprofit organizations are thus, to a significant extent, "mini-conglomerates," providing a wide range of services to their client groups.

While there is great diversity in the fields of nonprofit activity in Maryland, however, there are also some distinct patterns. Thus, as Table 3.1 shows:

- **Social Service Dominance.** The most common field of nonprofit action is *social services*, which includes everything from emergency assistance to family and individual counseling. Just over half of all Maryland agencies reported some involvement in some aspect of social services.

- **Education.** Just behind social services as a focus of nonprofit action is education, which engages nearly half of Maryland nonprofit organizations. Included here are actual educational institutions as well as parent-teacher associations and information centers of various sorts.

- **Advocacy and Community Development.** The third most commonly cited activity of Maryland nonprofit organizations, signifi-

Table 3.1
Fields of Activity of Maryland Nonprofit Agencies, 1996 (n=12,981)

Field	Agencies Reporting Some Activity in the Field	
	Number	Percent*
Social services	6,662	51%
Education	5,978	46
Advocacy, legal services	4,041	31
Community development	3,711	29
Culture and recreation	3,335	26
Philanthropy, volunteerism	2,516	19
Research	2,132	16
Environment, animal protection	2,131	16
Health	2,001	15
Mental health	1,908	15
Sports, recreation	1,784	14
Employment, job training	1,661	13
Crime, criminal justice	1,353	10
International	1,206	9

Source: MANO/JHU-IPS Maryland Nonprofit Survey, 1996

* Because agencies were free to report more than one service, percentages here total more than 100 percent.

cantly, is advocacy. Just over 30 percent of all agencies reported some involvement in advocacy, civil rights, or legal rights. In addition, 29 percent indicated some involvement in community development, which is very similar to advocacy. What this suggests is that a significant minority of Maryland nonprofit organizations remains committed to this fundamental function of the nonprofit sector and has not simply become absorbed in direct service functions.

Distribution of Agencies by *Primary* Field of Activity

Although most agencies are active in two or more fields, most nevertheless carry out the majority of their activities in a single field. In fact, if the "primary service field" is defined as the one in which an agency spends half or more of its total expenditures, then 85 percent of Maryland agencies can be assigned to a primary field. When this is done, the results are quite revealing.

The most common areas of concentration of Maryland nonprofit agencies are in the fields of arts and culture, social services, and education.

- **Three dominant fields.** In the first place, as Figure 3.1 shows, three fields--culture and recreation; social services; and education--clearly attract the majority of Maryland nonprofit agen-

cies. Taken together, these three components account for over half (56 percent) of all nonprofit agencies in the state. Not surprisingly, these are also the fields that involve the broadest segments of the population.

- **Two additional prominent fields.** In addition to these three largest fields, two other types of agencies absorb the energies of substantial shares of agencies--those engaged in housing, community development, and environmental protection; and those in the multiservice category. Close to 15 percent of all agencies fall into each of these two additional groupings.

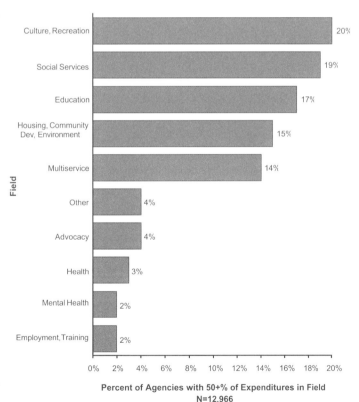

Figure 3.1 Distribution of Maryland Nonprofit Organizations, by Primary Field, 1996

Percent of Agencies with 50+% of Expenditures in Field
N=12,966

Source: MANU/JHU-IPS Maryland Nonprofit Survey, 1996

Size and Structure

Number of agencies, however, is only one measure of nonprofit activity. As Chapter Two showed, the distribution of nonprofit activity is quite different when the number of employees is used as the measure. The principal reason for this is that agencies differ considerably in size and scale of operation.

> *Two-thirds of Maryland nonprofit organizations have total expenditures of less than $25,000. Taken together, however, these agencies account for only 1 percent of total nonprofit expenditures in the state.*

Expenditures and Employment

These size differences are vividly apparent in terms of agency expenditures. Thus, as Figure 3.2 shows:

- Two-thirds of Maryland nonprofit organizations have total expenditures of less than $25,000. Taken together, however, these agencies account for only 1 percent of total nonprofit expenditures in the state.

- At the opposite extreme, 7 percent of the agencies have expenditures of $1 million or more. However, these agencies account for 93 percent of all expenditures.

- The Maryland nonprofit sector is thus composed of some rather disparate agencies, with the bulk of the resources in the hands of a small handful of agencies but the bulk of the agencies controlling only limited resources.

These disparities between the share of agencies and the share of resources can be found in virtually all service fields. In every field but one (mental health), for example, the majority of agencies is in the small category. And in every field but two (health and mental health), the proportion of agencies in the large category is less than 10 percent.

Despite wide differences in agency size within fields, however, agency size also varies sys-

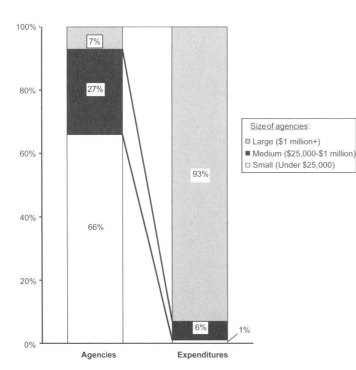

Figure 3.2 The Largest 7 Percent of Maryland Nonprofits Account for 93 Percent of Expenditures, 1994

Source: MANO/JHU-IPS Maryland Nonprofit Survey, 1996

Table 3.2
Average Agency Size, by Type of Agency, 1994

Type (n)	Average Expenditure per Agency
Advocacy (489)	$24,938
Culture, recreation (2,563)	181,957
Community development, environment (1,908)	261,663
Multipurpose (1,875)	288,619
Social Service (2,315)	311,226
Other (595)	601,394
Education (2,093)	908,032
Employment, training (211)	1,111,661
Mental health (295)	2,676,610
General health (371)	23,951,726
Average of All Agencies	**$1,133,055**

Source: MANO/JHU-IPS Maryland Nonprofit Survey, 1996

tematically among fields. Thus, as Table 3.2 shows:

- The average expenditure per agency for advocacy organizations is only $25,000 per year whereas the average for general health organizations is $24 million per year, or 1000 times as much. This is due to the presence among the health organizations not only of small clinics and professional associations but also of huge hospitals with hundreds of millions of dollars in annual expenditures.

- Generally speaking, as Table 3.2 shows, health, mental health, and employment and training organizations are the largest in terms of expenditures; and advocacy, cultural and recreation, community development, and social services organizations are the smallest.

Once we take account of these variations in agency size, the picture of nonprofit activity that emerges differs substantially from that shown in Figure 3.1. In particular, as Figure 3.3 shows:

Health organizations, though representing only 5 percent of all Maryland nonprofit agencies, account for the overwhelming majority (68 percent) of all nonprofit expenditures in the state.

- **Health Dominance.** Health organizations, though representing only 5 percent of all Maryland nonprofit agencies, account for the overwhelming majority (68 percent) of all nonprofit expenditures in the state. Included here, as the more detailed data in Table 3.3 indicate, are general health (hospitals, clinics), which account for 62 percent of the expenditures, and mental health facilities, which account for another 6 percent.

- **Human Services.** The second largest component of the Maryland nonprofit sector in expenditure terms is a variety of human service activities, including day care, personal social services, crisis intervention, community development, employment and training, and multipurpose agencies. These agencies tend, on average, to be much smaller than the health care providers, however. Though representing approximately half of all non-

Figure 3.3 Distribution of Nonprofit Expenditures, by Field, 1994

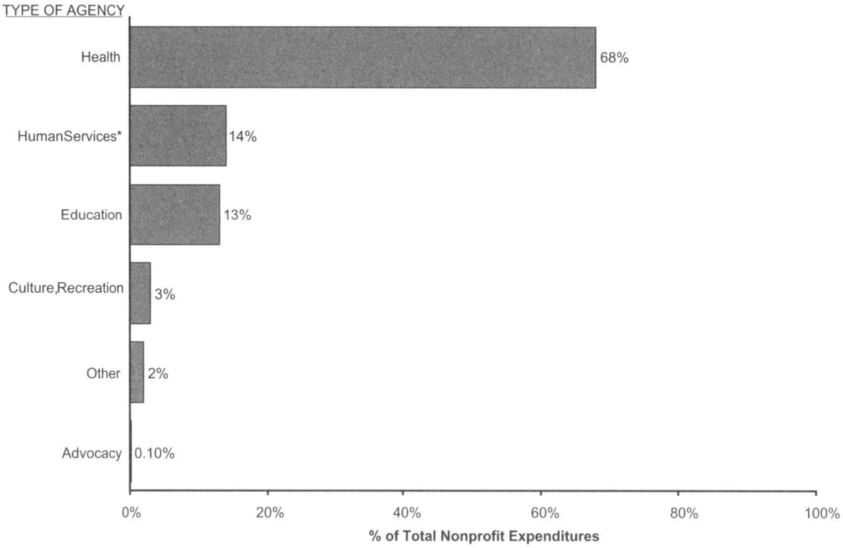

TYPE OF AGENCY

- Health: 68%
- HumanServices*: 14%
- Education: 13%
- Culture, Recreation: 3%
- Other: 2%
- Advocacy: 0.10%

% of Total Nonprofit Expenditures

Source: MANO/JHU-IPS Maryland Nonprofit Survey, 1996

* Includes social services, multiservice, community development, and employment and training organizations.

profit agencies in the state, they account collectively for only 14 percent of sector expenditures. Of this total, social service agencies account for 5 percent and multipurpose agencies for 4 percent.

- **Education.** The third largest component of the Maryland nonprofit sector in expenditure terms is education, including higher education, elementary and secondary education, and education support organizations. These organizations account for an estimated 13 percent of all nonprofit expenditures, compared to their 17 percent share of all agencies.

- **Limited expenditure shares elsewhere.** The remaining fields absorb a much smaller share of total nonprofit expenditures. Thus, while culture and recreation organizations comprise 20 percent of all organizations, they account for only 3 percent of the sector's expenditures. Similarly, organizations specializing in advocacy represent 4 percent of the agencies but bare-

Table 3.3
Distribution of Maryland Nonprofit Expenditures and Agencies, by Field, 1994

Field	Percent of Expenditures	Percent of Agencies
Health	**68%**	**5%**
General health	62	3
Mental health	6	2
Human services	**14**	**50**
Social services	5	19
Multipurpose	4	14
Community development	3	15
Employment and training	2	2
Education	**13**	**17**
Culture, recreation	**3**	**20**
Other	**2**	**4**
Advocacy	**-**	**4**
TOTAL	100%	100%

Source: MANO/JHU-IPS Maryland Nonprofit Survey, 1996.

Table 3.4
Distribution of Nonprofit Expenditures, by Field, Maryland and U.S., 1994

Field	% of Expenditures	
	Maryland	U.S.
Health	68%	61%
Human services	14	12
Education	13	22
Culture, Recreation	3	2
Other	2	3
TOTAL	100%	100%

Sources: Maryland data from MANO/JHU-IPS Maryland Nonprofit Survey, 1996; U.S. Data from *Nonprofit Almanac, 1996/97* (San Francisco: Jossey-Bass, 1996), p.190.

ly register in expenditure terms. This does not mean, of course, that nonprofits are unimportant in these fields. Rather, it means that non-profit activity in these fields is overshadowed in expenditure terms by the huge outlays of nonprofit hospitals, colleges and universities.

Comparison to National Averages

As we saw in Chapter Two with respect to employment, this overall structure of the nonprofit sector in Maryland is strikingly similar to that at the national level, though some differences are also apparent:

- Nationally as well, health dominates the nonprofit scene fiscally, though its position in Maryland appears even greater than the national average (68 percent of expenditures vs. 61 percent).

- The education component of the nonprofit sector, by contrast, is somewhat smaller in Maryland than in the nation at large (13 percent of expenditures vs. 22 percent). The results here may be due to sampling and return rate problems with the state's private higher education institutions, however.

- Elsewhere, the Maryland results appear to track the national picture remarkably closely.

Volunteer Usage

Before we rest content with this picture of nonprofit activity in Maryland, however, it is necessary to take account of the enormous volunteer involvement in this set of organizations. As we have seen, volunteers add an additional 85,000 workers to the full time equivalent workforce of the nonprofit sector. Four out of five agencies making use of volunteers relied on them, moreover, to provide direct services or operate agency programs. In addition, 62 percent of the agencies reported using volunteers for clerical and administrative support, and 58 percent for fundraising. To the extent that volunteers are distributed among agencies differently from paid staff or overall expenditures, our picture of the relative scale of different types of nonprofit activity could change drastically.

> *Small agencies make more extensive use of volunteers than they do of paid staff, yet the large agencies still make far greater use of volunteers than do the small agencies.*

To what extent is this the case? How does the distribution of volunteer input compare to the distribution of paid staff among Maryland nonprofit organizations? And how does the inclusion of volunteers alter the picture of nonprofit activity in the state that has been presented so far?

Part of the answer to these questions can be found in Table 3.5. As this table shows:

- Small agencies make more extensive use of volunteers than they do of paid staff (13 percent of volunteer time vs. 1 percent of all paid staff time).

Table 3.5
Distribution of Maryland Nonprofit Agencies, Employees, and Volunteers, by Size of Agency, 1994

Agency Size (Expenditures)	Percent of		
	Agencies	Employees[a]	Volunteers[a]
Small (Under $25,000)	66%	1%	13%
Medium ($25,000–$999,000)	27	12	44
Large ($1 million and over)	7	87	43
TOTAL	100%	100%	100%

Source: MANO/JHU-IPS Maryland Nonprofit Survey, 1996
[a] Based on full-time equivalent work week

7 percent of all agencies in the largest category (those with expenditures in excess of $1 million) absorbed 43 percent of the volunteers, and those in the medium category ($25,000 to $999,999 in expenditures) accounted for another 44 percent. Evidently, volunteers and paid staff are not simply substitutes for each other. To the contrary, effective recruitment and management of volunteers itself seems to require some professional, paid staff.

While the inclusion of volunteers does not fundamentally alter the overall picture of the structure of the nonprofit sector, it does change the picture somewhat since the use of volunteers apparently varies by field. Thus, as Figure 3.4 shows:

- At the same time, however, the large agencies make still greater use of volunteers than do the small agencies. Thus, while the 66 percent of all Maryland nonprofits that are small absorbed 13 percent of the full-time equivalent volunteers used by Maryland nonprofit agencies, the

- With volunteer activity included, the health share of overall nonprofit activity, while remaining in first place, drops sharply in rel-

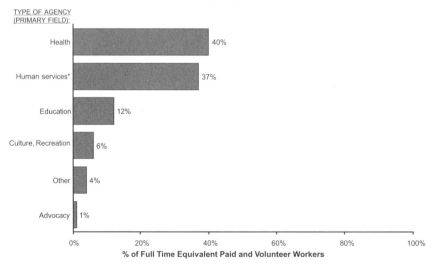

Figure 3.4 Distribution of Maryland Nonprofit Activity, Including Volunteers, by Major Field, 1994

Source: MANO/JHU-IPS Maryland Nonprofit Survey, 1996

* Includes social services, multiservice, community development, and employment and training organizations.

ative scale--from 68 percent of the total to 40 percent.

- Conversely, other human services, while remaining in second place, increases its overall share of total nonprofit activity from 14 percent with no volunteers included to 37 percent with volunteers. Of this total, as Table 3.6 shows, multiservice agencies account for 18 percent, and social service agencies for 12 percent. The large, nonprofit multiservice organizations thus absorb disproportionately large proportions of the volunteer effort in Maryland.

- Culture and recreation agencies also increase their relative share of overall nonprofit activity when volunteer effort is factored into the equation. Compared to 3 percent of the expenditures, such agencies absorb 6 percent of combined paid and volunteer effort.

- Finally, the inclusion of volunteer effort significantly boosts the advocacy involvement of nonprofit agencies since over half (55 percent) of the agencies using volunteers rely on them for advocacy activities.

In short, nonprofit agencies are active in a wide assortment of different fields in Maryland. In terms of expenditures and paid staff, the health component of this sector is clearly dominant. Once volunteer effort is considered, however, it is clear that health shares center stage with a great variety of other types of organizations providing a considerable range of human services. What is more, judging from agency reports about their use of volunteers, advocacy remains a vital component of nonprofit activity even though it does not dominate the expenditures of many agencies or the overall expenditures of the sector. What this suggests is that Maryland nonprofits are active both as providers of needed services and as supporters of civic involvement. They thus contribute to community well-being in both direct and indirect ways.

**Table 3.6
Distribution of Maryland Nonprofit Agency Activity, By Field, with Volunteers**

Field	Percent of Full Time Equivalent Workforce (w/volunteers) (n=259,000)
Health	40%
General health	34
Mental health	6
Human services	37
Multiservice	18
Social service	12
Community development	5
Employment and training	2
Education	12
Culture, recreation	6
Other	4
Advocacy	1
TOTAL	100%

Source: Author's computations based on MANO/JHU-IPS Maryland Nonprofit Survey, 1996

Once account is taken of volunteer input, the human service share of total nonprofit activity increases dramatically, from 14 percent with no volunteers included to 37 percent with volunteers.

II. AGENCY AGE

A Young Sector

If diversity is one of the key characteristics of the Maryland nonprofit sector, dynamism is a second. For all its historic importance, the Maryland nonprofit sector is composed mostly of very young organizations. In fact, as shown

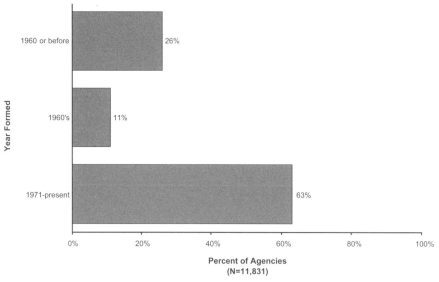

Figure 3.5 Distribution of Maryland Nonprofit Agencies by Year of Establishment

Source: MANO/JHU-IPS Maryland Nonprofit Survey 1996

in Figure 3.5, almost two-thirds of the organizations in existence at the time we conducted our survey in 1996 had been created in the previous 25 years, and over 40 percent in the previous fifteen. This suggests the critical role that the nonprofit sector performs as a mechanism for surfacing and responding to new social concerns and thus as a significant source of social vitality. It may also reflect the movement of nonprofit organizations to the suburbs that we identified in Chapter Two above since this has also necessitated considerable new agency formation.

Recent Growth Areas

The distribution of younger agencies is not uniform among different service areas, however. Rather, some fields have larger shares of young agencies than others. In particular, as Figure 3.6 shows:

Almost two thirds of the organizations in existence in 1996 had been created in the previous 25 years.

Advocacy is hardly a new role for Maryland's nonprofit organizations. Rather, some of the oldest agencies in the state are focused primarily on advocacy activities.

- Especially large shares of newer agencies are evident in the employment and training field, in social services, in community development and environmental protection, and in the "other" category. These reflect the increased attention to these fields in recent years. For example, day care has grown in importance as the proportion of women with small children who work has grown.

- By contrast, much higher proportions of multipurpose, advocacy, and education organizations were formed prior to 1961. Included here are the large family service agencies that have become so pivotal a part of the human service scene in the state.

Several conclusions seem to flow from these data:

- In the first place, the nonprofit sector does seem to be functioning as a mechanism for responding to new social concerns, such as environmental protection and the effort to equip unemployed people for work in the changing regional economy.

- In the second place, there is evidence here of the suburbanization trend identified earlier, especially in the social services field, which includes day care and residential care for the elderly. As people have moved increasingly to the suburbs, nonprofit organizations have had to be established as well. One reflection of this is that a disproportionately large share of the newer agencies is city or countywide, rather than neighborhood, focused. Even in the fields of health, social services, and culture and recreation, where significant numbers of large, pre-1961 agencies are in existence, substantial proportions of the existing agencies are relative newcomers. While this demonstrates the responsiveness of the non-profit sector, it also demonstrates the broad appeal of this sector, a point to which we will return below. Clearly, nonprofit organizations do not merely serve the poor. Rather, a significant part of the sector is involved in providing services to the suburbanizing middle class, and considerable portions of the charitable resources available are being absorbed in this pursuit.

- At the same time, it seems clear that some of the older multipurpose organizations have managed to adapt to the new suburbanization trend by extending their service areas and client focus and thereby capturing a meaningful portion of the changing "market" for their services.

- Finally, the data reported here make clear that advocacy is hardly a new role for Maryland's nonprofit organizations. Rather, some of the oldest agencies in the state are focused primarily on advocacy activities.

Older Agencies are Larger

These tentative conclusions find further

Figure 3.6 Percent of Maryland Nonprofit Agencies Formed Since 1980, by Field

Source: MANO/JHU-IPS Maryland Nonprofit Survey, 1996

Table 3.7
Share of Maryland Nonprofit Agencies and Expenditures, by Age of Agency, 1994

Year Formed	Share of Agencies	Share of Expenditures
1960 and earlier	26%	84%
1961-1970	11	4
1971-present	63	12
TOTAL	100%	100%

Source: MANO/JHU-IPS Maryland Nonprofit Survey, 1996

confirmation in data on the relationship between agency age and size. While the younger agencies are far more numerous than the older ones, they are also much smaller. Though less numerous, therefore, the older agencies still tend to dominate the state's nonprofit scene in fiscal terms. Thus, as Table 3.7 shows,

- The 63 percent of all Maryland nonprofit agencies that were formed since 1971 account for only 12 percent of the sector's expenditures.

- By contrast, the 26 percent of all agencies formed prior to the 1960s account for a striking 84 percent of the expenditures. This is not to say that all the older agencies are large or all the newer ones small. Nevertheless, the overall disparities in size are striking. Side by side with a large number of small and relatively new agencies stand a number of extremely large older agencies that provide the fiscal core of the state's nonprofit scene.

- Of special note here, moreover, are the large "multipurpose" organizations that operate in the state. Nearly half of these organizations were formed prior to the 1960s. They have evolved over time from single-purpose entities into sizable human service conglomerates offering a wide array of services to residents both in the urban core and in the surrounding areas. In the process, they provide an important ballast to the state's nonprofit sector.

The 26 percent of all agencies formed prior to the 1960s account for a striking 84 percent of the state's nonprofit organization expenditures.

In short, we can now add to our portrait of Maryland's nonprofit sector the realization that this sector contains some very different institutions. On the one hand, the sector contains a group of larger, older agencies in most of the key fields, many of them evolving over time from a particular activity into essentially "conglomerates" that are capable of responding in an integrated way to a wide variety of human problems. At the same time, the sector displays a significant degree of openness to new entrants and approaches. In this way the nonprofit sector offers both a high degree of stability and permanence, and a high degree of flexibility and responsiveness to emerging needs.

III. WHOM DOES THE MARYLAND NONPROFIT SECTOR SERVE?

But what are the needs to which the Maryland nonprofit sector responds? More specifically, whom do these organizations serve? Who are the beneficiaries of their activities? And how does this vary by type of organization? The previous discussion has already suggested some tentative answers to these questions, but it is now time to address this issue more directly.

Doing so is by no means an easy task, however. It is often not clear who actually benefits from a particular service of a nonprofit agency, as the client may not always be the intended target of the program. For example, a prenatal nutrition program serves the mother but the underlying goal of the program is to improve infant health. Furthermore, definitions of clientele or target population vary greatly, even among agencies providing similar services. One day care program may consider its clients to be children, another parents, a third the mothers, and a fourth the families of the children. A neighborhood legal services clinic may define individual clients as its service population while another may define it in terms of the residents of the neighborhood in which the clinic operates. One arts group may count its paid attendance as the patrons served while another measures its patron base in terms of the size of the potential theater-going community.

Despite these difficulties, our survey enables us to shed some light on the question of the clientele and beneficiaries of Maryland's nonprofit agencies.

> *The argument crediting nonprofit organizations with a significant community role finds considerable support in the Maryland data.*

In particular, we asked questions about the geographic focus of agency operations and about agency clientele defined in terms of age, race, ethnicity, and social and economic status. The results suggest the breadth of nonprofit involvement in Maryland life.

Geographic Focus

One of the strengths traditionally claimed for the nonprofit sector is its responsiveness to particular neighborhood concerns. Nonprofit organizations are consequently perceived as an antidote to the impersonality of much of modern life, a way to give a sense of belonging to people alienated from the larger mega-structures of modern society--the large corporations, governmental bureaucracies, and even professional organizations.[1] To what extent does this argument apply to the Maryland nonprofit scene?

Figure 3.7 provides a first approximation of the answer to this question by documenting the geographic focus of Maryland's nonprofit agencies.

Figure 3.7 Geographic Service Area of Maryland Nonprofits, 1996

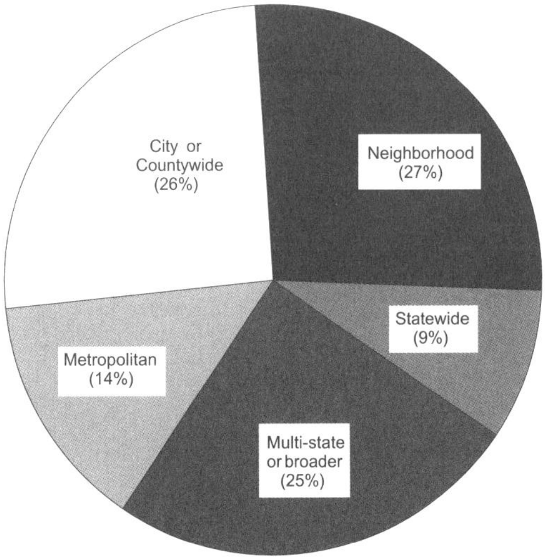

- Neighborhood (27%)
- City or Countywide (26%)
- Multi-state or broader (25%)
- Metropolitan (14%)
- Statewide (9%)

Source: MANO/JHU-IPS Maryland Nonprofit Survey, 1996

---

[1] This argument was developed most forcefully by Robert Nisbet in *Community and Power*, 2nd edition. (New York: Oxford University Press, 1962). It has been restated recently in Robert Putnam, *Making Democracy Work: Civic Traditions in Modern Italy* (Princeton: Princeton University Press, 1993).

- **Neighborhood Focus.** As it turns out, the argument crediting nonprofit organizations with a significant community role finds considerable, though by no means overwhelming, support in the Maryland data. Thus the largest single group of agencies, accounting for 27 percent of the total, report a neighborhood focus, and another 26 percent operate primarily at the city or county level. This is particularly true, moreover, in the Baltimore region, where 34 percent of the agencies indicate a neighborhood focus. The sizable presence of neighborhood associations in Baltimore City very likely explains this, at least in part. Apparently, providing an institutional focus for neighborhood concerns is one of the functions that Maryland nonprofit organizations perform.

> *As the nonprofit sector has "suburbanized", it has apparently surrendered some of its special neighborhood character and appeal.*

- **Metropolitan Agencies.** While a significant share of Maryland's nonprofit organizations are essentially local in focus, however, a considerable portion also have a broader orientation. Thus, 22 percent of the organizations serve a metropolitan or statewide clientele. They therefore provide a mechanism for addressing problems that span local governmental jurisdictions.

- **Multi-state and International Agencies.** In addition, 25 percent operate at a multi-state and even international level. This doubtless reflects the state's proximity to Washington, D.C., the hub of national and international political life. In fact, the proportion of national and international agencies among the agencies in Maryland's Washington suburbs is 27 percent. But the Baltimore region has also attracted numerous nonprofit agencies that want easy access to Washington without having to pay Washington prices.

- **Recent Trends.** Finally, and perhaps most significantly, the geographic focus of agency activities has changed quite extensively over time. In particular, the older agencies are much more likely to have a neighborhood focus than the newer agencies. Thus, as Table 3.8 shows, over 40 percent of the agencies formed in the

Table 3.8
Geographic Focus of Maryland Nonprofit Agencies, by Age of Agencies

Geographic Focus	All	1960 or earlier	1961-70	1971-90	Since 1990
Neighborhood	27%	42%	45%	19%	13%
City or Countywide	26	16	12	33	37
Metropolitan or Statewide	22	24	18	18	34
Multistate or broader	25	18	25	30	16
TOTAL	100%	100%	100%	100%	100%

Source: MANO/JHU-IPS Maryland Nonprofit Survey, 1996
Calculations may be subject to rounding error

1960s and earlier had a neighborhood focus compared to less than 20 percent of the agencies formed since then. As the nonprofit sector has "suburbanized," in other words, it has apparently surrendered some of its special neighborhood character and appeal.

Clientele and Beneficiaries

If geographic focus is one indication of who benefits from Maryland nonprofit agencies, client focus is another. Nonprofit organizations have often been touted as a more personal mechanism for responding to human needs, one that respects the ethnic, religious, and other social distinctions that give special meaning to people's lives. Others think of the nonprofit sector as an essentially "charitable" set of institutions in the dictionary definition of that term-- i.e. providing help primarily to the poor and disadvantaged. To what extent do our survey results support these expectations?

The central conclusion that emerges from the data is that agencies do not seem to target particular client groups as much as we might expect.

To answer these questions, we asked agencies to report on the characteristics of those who use their programs or services defined in terms of ethnicity, race, economic circumstance, age, and other characteristics. The central conclusion that emerges from the data is that agencies do not seem to target particular client groups as much as we might expect. Rather, most agencies seem to serve a fairly diverse clientele. Thus, as Figure 3.8 shows:

• **Limited Poverty Focus**. For only 16 percent of Maryland nonprofit agencies do the poor comprise the majority of the clients. Even when we relax the definition of poverty focus, moreover, this general picture does not change much. Thus, as Figure 3.8 also indicates, the share of agencies with even 10 percent of their clients in poverty is only 26 percent. In other

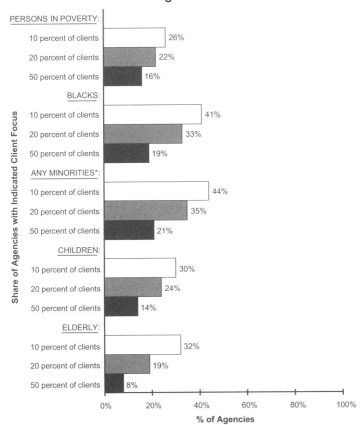

Figure 3.8 Client Focus of Maryland Nonprofit Agencies

PERSONS IN POVERTY:
- 10 percent of clients: 26%
- 20 percent of clients: 22%
- 50 percent of clients: 16%

BLACKS:
- 10 percent of clients: 41%
- 20 percent of clients: 33%
- 50 percent of clients: 19%

ANY MINORITIES*:
- 10 percent of clients: 44%
- 20 percent of clients: 35%
- 50 percent of clients: 21%

CHILDREN:
- 10 percent of clients: 30%
- 20 percent of clients: 24%
- 50 percent of clients: 14%

ELDERLY:
- 10 percent of clients: 32%
- 20 percent of clients: 19%
- 50 percent of clients: 8%

Source: MANO/JHU-IPS Maryland Nonprofit Survey, 1996
* Blacks, Hispanics, or Asians

words, for the vast majority of Maryland nonprofit agencies, the poor do not comprise even 10 percent of their clients. Quite clearly, Maryland's nonprofit sector is not primarily "charitable" in this narrow sense of the word. Rather, the vast majority of agencies serve mostly middle class clients rather than focusing primarily on the poor.

> *Maryland's nonprofit sector is not primarily "charitable" in the narrow sense of the word. Rather, the vast majority of agencies serve mostly middle class clients rather than focusing primarily on the poor.*

- **Limited Racial or Ethnic Focus.** A similar picture emerges with respect to racial or ethnic focus. Thus, as Figure 3.8 also indicates, Blacks comprise the majority of clients of only 19 percent of Maryland nonprofit agencies; and *any minorities*, including Blacks as well as Hispanics and Asians, comprise the majority of clients of only 21 percent of the agencies. There is thus little evidence here that the nonprofit sector is functioning as a mechanism for group identity, at least to the extent that groups are defined in racial or ethnic terms.

- **Limited Age Focus.** Nor do Maryland nonprofit agencies tend to focus on particular clientele defined in age categories. Thus, children are the primary focus of only 14 percent of the agencies and the elderly the primary focus of only 8 percent.

Poor-Serving Agencies: Closer Look

Given the importance of the presumed "charitable" character of nonprofit organizations, it is important to examine this facet of agency operations in a bit more detail. How can we make sense of the finding that so few Maryland agencies focus primarily, or even modestly, on the poor? Is this a function of the broad range of agencies that are included within the purview of this sec-

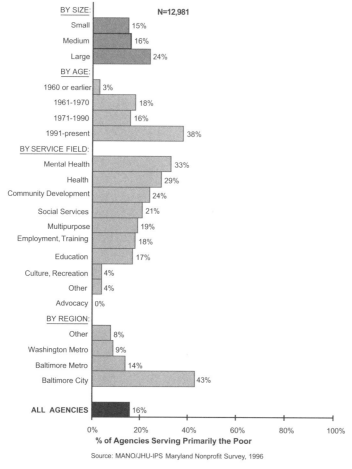

Figure 3.9 Characteristics of Primarily Poor Serving Maryland Nonprofit Agencies

N=12,981

BY SIZE:
- Small: 15%
- Medium: 16%
- Large: 24%

BY AGE:
- 1960 or earlier: 3%
- 1961-1970: 18%
- 1971-1990: 16%
- 1991-present: 38%

BY SERVICE FIELD:
- Mental Health: 33%
- Health: 29%
- Community Development: 24%
- Social Services: 21%
- Multipurpose: 19%
- Employment, Training: 18%
- Education: 17%
- Culture, Recreation: 4%
- Other: 4%
- Advocacy: 0%

BY REGION:
- Other: 8%
- Washington Metro: 9%
- Baltimore Metro: 14%
- Baltimore City: 43%

ALL AGENCIES: 16%

% of Agencies Serving Primarily the Poor

Source: MANO/JHU-IPS Maryland Nonprofit Survey, 1996

tor? Are there variations among agencies in terms of their poverty focus? If so, which agencies tend to focus most heavily on the poor and which the least?

Figure 3.9 reports the answers our data provide to these questions. As this figure indicates:

- **Size and Poverty Focus.** In the first place, large agencies tend to target the poor more heavily than small and medium-sized agencies. One reason for this finding may be that larger agencies tend to receive government support—a point to which we will return in the next chapter. Because government programs are often targeted on the poor, the heavier poverty focus of the larger agencies may reflect the impact of government funding priorities on agency behavior.

- **Agency Age and Poverty Focus.** Younger agencies also tend to focus more heavily on the poor than do older agencies. What this suggests is that as agencies age they extend their efforts to broader client groups. One reason for this may be that the agencies' original clients improve their economic circumstances yet remain involved with the agency. This is the case, for example, with agencies that began as vehicles for self-help among ethnic immigrants.

At least as plausible an explanation, however, is that the agencies that survive are those that find a market niche that can sustain them over the long run, and this requires some broadening of their appeal to clients who can afford the services they offer. At the same time, one of the more heartening tendencies evident in our data is the apparent emergence of numerous newer agencies serving primarily poor people. Compared to 16 percent of all agencies, nearly 40 percent of the agencies formed during the 1990s report focusing primarily on the poor.

- **Poverty Focus and Field of Service.** Considerable variations also exist among agencies of different types in the extent of their attention to the poor. Interestingly, the agencies with the heaviest concentration on the poor are in the health field, particularly mental health. One third of all mental health agencies and nearly 30 percent of all health providers report that the poor comprise half or more of their clients. Community development agencies also display a fairly high concentration on the poor: one-fourth of these agencies primarily serve the poor.

Interestingly, however, only about one in five of the social service and multipurpose agencies focus primarily on the poor. This is consistent with national data but is still somewhat surprising since these fields are commonly considered to be heavily oriented to the disadvantaged. Evidently, nonprofit social service providers have broadened their focus to clientele beyond the poor, especially as they have moved to the suburbs. This is certainly true, for example, of day care centers and facilities for the elderly, many of which serve middle class clients who can pay fees for their services.

> *Over 40 percent of Baltimore City agencies report focusing primarily on the poor. By contrast, in the other regions of the state the proportion is barely one-fourth as much.*

> *One of the more heartening tendencies evident in our data is the apparent emergence of numerous newer agencies primarily serving poor people.*

- **Location and Poverty Focus.** The tendency of the nonprofit sector to lose its poverty focus as it "suburbanizes" is even more dramatically apparent in Figure 3.9's data on the variation of poverty focus among agencies in different parts of the state. Not surprisingly, perhaps, the agencies focusing primarily on the poor are heavily concentrated in Baltimore City. Over 40 percent of Baltimore City agencies report focusing primarily on the poor. By contrast, in the other regions of the state the proportion is barely one-fourth as much. The state's nonprofit sector is thus bifurcated, with a large poverty-focused segment in Baltimore City co-existing with a segment that focuses much more heavily on the broad middle class elsewhere in the state. In this sense the nonprofit sector reflects, as well as tries to alter, the prevailing social and economic realities of the state.

Conclusion

The nonprofit sector is thus a diverse and dynamic component of Maryland life. Not only is it an economic force, but also it is a unique community resource, providing an immense array of services to a broad cross-section of Maryland citizens. What is more, this is a vibrant sector that provides a flexible way to respond to human needs as they arise without having to wait for the cumbersome processes of consensus-building and mass education required to mobilize governmental support.

At the same time, some puzzling anomalies are also apparent in the data reviewed here. One of the most important is the relatively limited tendency of Maryland nonprofit agencies to focus their attentions on the poor. While agencies located in Baltimore City do focus heavily on the poor, the same is not true to anywhere near the same extent for agencies operating elsewhere in the state; and even in Baltimore City less than half of all the agencies report focusing primarily on poor people.

One reason for the relatively limited poverty focus of most agencies is very likely the challenge these organizations face in generating financial support. To understand some of the features of the nonprofit sector identified here, therefore, it is necessary to turn from this analysis of agency characteristics to an examination of the financial realities that lie behind them.

CHAPTER FOUR
NONPROFIT FINANCES

Nonprofit organizations are distinguished from other corporations in the private sector by their relative freedom from the "bottom line," from the need to generate profits. But this hardly means that these organizations are not in need of resources. To some extent these resources are non-financial in character. They take the form of in-kind contributions of time and products. But nonprofit organizations are also employers and purchasers of a variety of goods and services. As such, they are deeply enmeshed in the market economy and must consequently generate revenues to cover their costs.

What is the source of these revenues? Where do nonprofit organizations get the resources they need to operate? How does this vary by type or size of organization and by region? What effect do funding patterns have on the client focus of agencies? What is the relative scale of financial and non-financial resources that nonprofit organizations use?

The purpose of this chapter is to answer these questions for the Maryland nonprofit sector. To do so, the discussion focuses first on cash income. It then broadens the lens to embrace volunteer inputs as well.

Private giving from all sources--individuals, foundations, corporations--comprises less than 5 percent of total nonprofit sector income in Maryland.

I. OVERVIEW OF MARYLAND NONPROFIT FINANCES

According to popular conceptions, the principal source of support for nonprofit activity is private philanthropy, i.e., contributions from individuals, foundations, and corporations. This conception is fueled by regular celebrations of the scale of private charitable giving in the United States and by the almost endless fundraising campaigns that seem to be underway in this and other communities.

Limited Role of Private Philanthropy

In reality, however, these popular conceptions are quite wide of the mark. Not only is private philanthropy only one source of nonprofit revenue, but also it is far from the most important source. To the contrary, as Figure 4.1 shows:

- Private giving from all sources--individuals, foundations, corporations--comprises only about 4 percent of total nonprofit sector income in Maryland.

- This pattern of nonprofit finance is certainly not unique to Maryland. Nationally as well private giving accounts for a relatively small share of total nonprofit income. At the same time, private philanthropy's share of total

Figure 4.1 Major Revenue Sources of Maryland Nonprofit Agencies, 1994

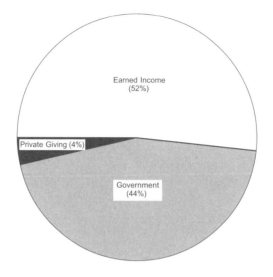

Source: MANO/JHU-IPS Maryland Nonprofit, Survey 1996

nonprofit income is *proportionally two and one-half times smaller in Maryland than it is nationally.* Compared to its 4 percent share of total nonprofit revenue in Maryland, private giving accounts for closer to 10 percent of nonprofit income nationally.[1]

- This situation is true despite the fact that private giving is the *most common* type of nonprofit income. As Table 4.1 shows, a higher proportion of nonprofit agencies receive some income from at least one source of private giving than from any other source. Thus 77.5 percent of agencies reported some income from private giving, most of them (68.5 percent) from private individuals.

- The other sources of private giving--foundations, corporations, and federated funders--reach far fewer agencies and account for even smaller shares of total nonprofit income. Thus only 25 percent of the organizations benefit from corporate support, and this support accounts for only 0.3 percent of total nonprofit income. Foundations provide a slightly larger 0.7 percent of nonprofit income in the state but touch a smaller share of agencies (18 percent). Finally, United Way and other federated funders have even more restricted reaches and contribute, collectively, only 0.7 percent of total income. In each of these areas, moreover, Maryland ranks far behind the national averages. Evidently, the philanthropic sector is less well developed and private philanthropy more limited in this state than is the case nationally.

Private philanthropy's share of total nonprofit income is proportionally two and one-half times smaller in Maryland than it is nationally

Earned Income the Major Source

Instead of private giving, the major source of support for Maryland nonprofit agencies, as Figure 4.1 shows, is earned income from fees, related businesses, unrelated businesses, and investments. Such earned income comprises over 50 percent of the total revenue of

---

[1] Based on data in *Nonprofit Almanac: Dimensions of the Independent Sector* (San Francisco: Jossey Bass Publishers, 1996), p. 190.

Table 4.1
Sources of Support of Maryland Nonprofit Organizations

Source	Share of Income from Source	Share of Agencies with Any Income From Source
Private Giving	4.4%	77.5%
Individuals	2.2	68.5
Corporations	0.3	25.3
Foundations	0.7	17.8
United Way	0.3	12.3
Other Federated Funders	0.4	4.6
Unallocated	0.5	N.A.
Government	43.5%	36.3%
Earned Income	52.1%	70.3%
Fees, charges	45.9	47.4
Related business income	2.9	10.9
Unrelated business income	0.5	3.6
Investment income	1.4	32.0
Special events	0.8	27.1
Other	0.6	16.1
Total	**100.0%**	

Source: MANO/JHU-IPS Maryland Nonprofit Survey, 1996
Calculations may be subject to rounding error
N.A. = Not Available

Maryland nonprofit agencies, and almost as many agencies receive income from this source (70 percent) as from private giving.

As Table 4.1 shows:

- The largest of the sources of earned income by far is *fees and service charges* that agencies receive from clients for the services they render. A sizable 46 percent of all Maryland nonprofit income comes from this one source, and nearly half of all agencies benefit from it. This is consistent with our earlier finding that Maryland nonprofit agencies serve a broad cross-section of clients and hardly restrict their attention to those who are needy. Evidently, many of these clients pay for the services they receive.

Instead of private giving, the major source of support for Maryland nonprofit agencies is earned income from fees, related businesses, and investments. This source accounts for over half of all nonprofit income in the state.

- Of the agencies charging fees for their services, however, most do so for only a portion of their clientele, and even then, most cover only a portion of the costs. Thus, only 14 percent of the agencies reported that they charge some kind of fee for *all* of their clients, and only 25 percent reported that they do so for *three-fourths or more* of their clients. Evidently, most agencies charging fees restrict them to clients who are able to pay. Similarly, of those charging fees, only 11 percent indicate that they cover the full cost of the affected services through fees, and only 21 percent indicate that they cover as much as three-fourths of the costs. What this suggests is that agencies utilize fees essentially to supplement their other revenues, and even then rely on the fees they are able to collect from some clients to cross-subsidize services for other clients. Nevertheless, the scale of fee income remains immense.

- Other sources of earned income are much less significant in either scale or reach. Thus, for example, 32 percent of all Maryland nonprofit organizations reported receiving some income from *investments*, but investment income accounts for just over 1 percent of total income.

- *Related business income*, i.e. income from business activity that is related to the mission of the agency (e.g. sales of reproductions by museums), is somewhat larger in scale, but still accounts for only 3 percent of total income and affects only 11 percent of all agencies. Interestingly, however, this still means that related business income provides a larger share of the total income of Maryland nonprofit agencies than individual giving.

- *Unrelated business income*, i.e. income from businesses that are unrelated to the organizations' charitable missions, is far less significant despite the attention it receives in press accounts. Based on our survey responses, less than 1 percent of nonprofit income in Maryland comes from this source and less than 5 percent of all agencies have access to it.

The Prominence of Government Support

If earned income is the major source of nonprofit income in Maryland, government support is a close second. In fact:

- Federal, state, and local governments provide 44 percent of the income of Maryland nonprofit agencies, or more than two out of every five dollars. Clearly, government has turned extensively to nonprofit organizations to carry out public objectives in Maryland.

- This pattern of government support to nonprofit agencies is certainly not unique to Maryland. However, it is significantly more prevalent here. Compared to the 44 percent of total nonprofit income that comes from government in Maryland, such support accounts for only about 34 percent of nonprofit income nationally.[2] Clearly, Maryland nonprofit agencies have been particularly effective in generating governmental support for their activities, and government agencies have turned more extensively to nonprofit organizations to help them carry out their missions than is the case nationally. Government-nonprofit partnership is thus a special cornerstone of nonprofit action in Maryland.

- Government support also touches a significant share of Maryland nonprofit agencies. Well over a third of all agencies receive such support, making this the third most common source of income, behind only individual giving and service fees, but ahead of foundation, corporate, and United Way support as well as related business income.

> *Government-nonprofit partnership is a special cornerstone of nonprofit action in Maryland.*

Summary

In short, the pattern of nonprofit finance in Maryland differs dramatically from what popular conceptions would lead us to expect. For one thing, private giving ranks far lower than most people believe. In fact, it ranks far lower than what even national figures would suggest. For whatever reason, the Maryland nonprofit sector, though relatively as large as that in other states, does not seem to enjoy even the limited levels of private support evident in other parts of the nation. One reason for this may be the lack of "head offices" of major corporations in Maryland, which limits corporate philanthropic involvement. Another is the generally limited foundation presence and relatively low priority placed on nonprofit and philanthropic matters within the corporate and philanthropic community. Finally, Maryland's citizens may be more accustomed to looking to government to solve major social welfare and urban development problems. The state has a significant statewide income tax that is used to finance local as well as state government and citizens may consequently feel less inclined to support charitable purposes. Whatever the reason, the limited scale of private giving is one of the striking features of the Maryland nonprofit scene.

---

[2] Computed from data in *Nonprofit Almanac, 1996/97*, p. 190.

That Maryland nonprofit organizations have been able to grow to the extent they have in the face of this relatively limited private support has been due in large part to their success in generating earned income and tapping government support. While earned income is the larger of the two, it is government support that distinguishes the Maryland nonprofit sector from its counterparts elsewhere. Maryland agencies have been unusually successful in forging partnerships with the public sector to deliver a wide range of human services and respond to community needs.

Far from displacing the nonprofit sector, government has powered its growth in Maryland, enabling nonprofits to extend their reach and strengthen their role. At the same time, however, this past success makes the Maryland nonprofit sector unusually vulnerable to the kinds of budget cuts that have recently figured so prominently in national political life.

II. PATTERNS OF NONPROFIT FINANCE

The picture of nonprofit finances painted so far treats all nonprofit organizations as a group. Given the diversity of Maryland's nonprofit sector, however, it is at least possible that this aggregate picture may obscure as much as it reveals, that the revenue structure for some types of agencies may differ markedly from that for others.

To what extent is this the case? What variations exist in the financing of nonprofit agencies of different sizes, ages, and fields?

As Figure 4.2 shows, the answer to this question is somewhat ambiguous. On the one hand, significant variations are apparent in the funding patterns of different types of organizations. On the other hand, however, the general pattern that is evident in the aggregate results is also widely apparent at the subsector level.

Variations by Field

In the first place, the top two sources of income for the Maryland nonprofit sector as a whole--government and earned income-- also turn out to be the top two sources of support for most of the individual subsectors. Thus:

The Maryland nonprofit sector, though relatively as large as that in other states, does not seem to enjoy even the limited levels of private support evident in other parts of the nation.

- Of the ten types of agencies we identified in our survey, five get the preponderance of their income from government support. Included here are mental health, employment and training, social services, health, and multipurpose agencies. Clearly, government support is quite important to a broad cross-section of Maryland's nonprofit sector.

- Of the five remaining types of agencies, four get the majority of their support from earned income. Included here are community development, education, culture and recreation, and other agencies. In addition, earned income accounts for more than 49 percent of total income in the huge field of health, where it shares top funding honors with government.

- In only one field--advocacy--does private giving provide the majority of funding, and this is a relatively small, though important, field of nonprofit action.

- While private giving is the dominant source of income in only one field, however, it nevertheless plays a considerable role in the funding of two other types of agencies as well--social service and multipurpose. In both of these fields over 20 percent of the revenue comes from private charitable giving. Nevertheless, even when we exclude the two very large fields where the role of private giving is limited--health and education--the private giving share of total nonprofit revenue in Maryland still stands at only 10 percent, which is well below the comparable national figure.[3]

Variations by Size of Agency

A considerably different picture of the funding of nonprofit organizations in Maryland emerges, however, when we take account of agency size. As Figure 4.2 shows:

- Earned income and government are the dominant sources of income of the large and medium sized agencies, respectively.

- Among the small agencies, however, private giving emerges as the major source of support. Nearly half of the income of these agencies comes from private giving.

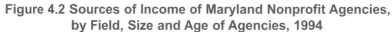
The top two sources of income for the Maryland nonprofit sector as a whole-- government and earned income--also turn out to be the top two sources of support for most of the individual subsectors.

- Because the smaller agencies are far more numerous than the larger ones, moreover, this means that a substantial proportion of nonprofit agencies in the state receive the preponderance of their income from private giving even though private giving constitutes a small share of the aggregate income of the sector. This point is evident in Figure 4.3, which records the share of agencies reporting private giving, earned income, and government, respectively, as their *principal* source of income. As this figure shows:

- Earned income turns out to be not only the largest aggregate source of support for Maryland nonprofit agencies, but also the one that is predominant for the largest share of agencies. Thus 42.3 percent of all Maryland nonprofits get the preponderance of their income from this source.

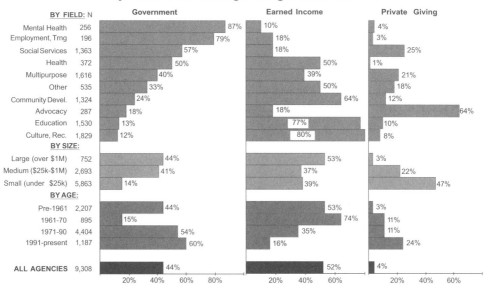

Figure 4.2 Sources of Income of Maryland Nonprofit Agencies, by Field, Size and Age of Agencies, 1994

Source: MANO/JHU-IPS Maryland Nonprofit Survey, 1996

---

[3]For the nation as a whole, private giving constitutes 17.4 percent of the income of all nonprofit service organizations other than health and education. Computed from *Nonprofit Almanac, 1996/97*, p. 190.

- Significantly, however, the source that provides the preponderance of support for the second largest number of agencies is private giving. Just over 40 percent of all Maryland agencies get the preponderance of their income from this source. In other words, while hardly the dominant revenue source in dollar terms, private giving is nearly the most common major source in terms of number of agencies relying chiefly on it for their support.

- By contrast, government support, though accounting overall for more than 40 percent of total nonprofit revenue, turns out to be the dominant source of support for a relatively small share of the agencies (16.5 percent), though many recipients of government support also receive significant shares of their income from fees and charges.

What this suggests is that private giving, though smaller in scale, may still be playing a distinctive function in the Maryland nonprofit world, a function akin to that played by "venture capitalists" in the business sector--i.e. supporting start-up activity and thus contributing to the responsiveness of the nonprofit sector to new societal needs. At the same time, if private support helps with the start-up of new initiatives, it is government support that helps to translate promising beginnings into full-fledged programs. This finding is consistent with what has come to be known as the "partnership theory" of nonprofit action, which views government and private philanthropy as partners rather than competitors in the operation of nonprofit organizations.[4]

Among the small agencies, private giving emerges as the major source of support.

Variations by Age of Agency

This interpretation of the relative roles of private giving and government support in the operation of the Maryland nonprofit sector

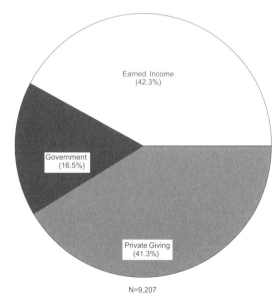

Figure 4.3 Percent of Maryland Nonprofit Agencies, with Indicated Principal Sources of Income, 1994

N=9,207

Source: MANO/JHU-IPS Maryland Nonprofit Survey, 1996

---

[4] For further elaboration of this theory, see Lester M. Salamon, "Of Market Failure, Government Failure, and Third Party Government: Toward a Theory of Government-Nonprofit Relations in the Modern Welfare State," in Lester M. Salamon, *Partners in Public Service* (Baltimore: The Johns Hopkins University Press, 1995), pp. 33-49.

finds further support in data on the variation in funding by agency age. Thus, as Figure 4.2 shows:

- Private giving is disproportionately important in the funding of young agencies. Compared to the 4 percent of total nonprofit income that comes from private giving, 24 percent of the income of young agencies originates with private philanthropy. Reflecting this, private giving is the predominant source of support for 54 percent of the younger agencies.

- While private giving is disproportionately important to the financing of start-up activity in the Maryland nonprofit sector, however, it does not have a monopoly on this function. To the contrary, government also plays a major role. In fact, it plays *the* major role in financial terms. Indeed, the younger the agencies, the larger the government share of their total revenue. Thus, 60 percent of the income of the youngest agencies, and 54 percent of the income of the next youngest agencies, comes from government. This confirms again the important "enabling" function that government performs for nonprofit organizations in Maryland, allowing new nonprofits not only to form but to begin making meaningful contributions to the alleviation of community problems. Indeed, with private philanthropy limited in scale and heavily committed to existing agencies, it is ironically government that has really helped to keep nonprofit organizations on the forefront of social problem solving.

- Government support is not only important to the youngest agencies, however. It also plays a significant role in the funding of older agencies. Thus, 44 percent of the income of agencies formed prior to 1961 comes from government sources. What this makes clear is that the expansion of government involvement in the human service field in the 1960s and 1970s did not only lead to the formation of new agencies organized to capture the new government funds, but also aided many older agencies as well. Thanks to the expansion of government support, these agencies were enabled to perform tasks, and very likely serve clientele, that were previously beyond their reach because of the financial limitations they faced. Government aid thus seems to have been "empowering" for the entire Maryland nonprofit sector.

Client Focus

Further confirmation of the idea that government support enabled the Maryland nonprofit sector to extend its reach can be found in data relating the funding structure of agencies to their propensity to serve the poor. Thus, as Table 4.2 shows:

Government aid seems to have been "empowering" for the Maryland nonprofit sector, enabling it to take on tasks, and serve clientele, that were previously beyond its reach.

- Agencies that primarily serve the poor tend to have much higher levels of government support than those that do not. Indeed, among primarily poor-serving agencies, government accounts for 74 percent of total income. By contrast, among agencies with fewer than 50 percent poor clients government support is 42 percent.

- Private giving also seems to play a special role in supporting agencies that serve the poor, though this role is still rather modest. Thus, compared to the 4 percent of overall nonprofit revenue that it provides, private giving accounts for 11 percent of the income of agencies primarily serving the poor.

- By contrast, reliance on earned income appears to be highly incompatible with service to the poor, despite the tendency of agencies with fee income to utilize sliding fee scales. Thus, the greater the reliance on earned income, the less likely an agency is to focus heavily on the poor. In particular, among agencies that primarily serve the poor, only 15 percent of total revenue comes from earned income.

By contrast, among other agencies, 55 percent of total revenue comes from earned income. This finding is highly significant in view of recent evidence suggesting that reductions in government support are likely to lead nonprofit organizations to rely more heavily on earned income. From the data presented here, it appears that the major losers in this shift will be poor people.

Regional Variations

One final interesting source of variation in the financing of nonprofit action in Maryland is regional in character. In particular, as Figure 4.4 shows, while there is a considerable amount of consistency among the various regions in terms of the shares of income coming from the various sources, one region seems to be an outlier. In Maryland's Washington suburbs, nonprofit organizations rely far more heavily on earned income and far less heavily on government support than is the case anywhere else in the state. The reason for this is very likely the distinctive composition of the nonprofit sector in this region. As Chapter Two showed, a significantly larger share of the nonprofit activity in this region is civic and social in character than is the case in the state as a whole. Social and recreational clubs are far more likely to be member supported than is the case for other types of agencies. In addition, given the population composition of these counties, it is likely that even their social service agencies receive considerable fee income for such services as day care and elderly care.

Implications

What emerges from this analysis is a picture of complementary funding sources financing nonprofit action in Maryland. This picture differs considerably from the image of conflict and competition that is frequently portrayed among ideologues in the field.[5] According to these accounts, a conflict exists between reliance on the public sector and reliance on the private sector in supporting nonprofit activity, with the latter clearly preferable to the former. In fact, what the data here suggest is that these two sources work hand-in-glove, with private philanthropy helping to ferret out

> *Agencies that primarily serve the poor tend to have much higher levels of government support than those that do not. This confirms the "enabling" role that government plays for nonprofit agencies.*

Table 4.2
Sources of Income of Maryland Nonprofit Agencies, by Client Focus

Source	Primarily Poor-Serving	Not Primarily Poor-Serving	All
Government	73.7%	41.6%	43.5%
Private Giving	11.4	3.9	4.4
Earned Income	14.9	54.5	52.1
TOTAL	100.0%	100.0%	100.0%

Source: MANO/JHU-IPS Maryland Nonprofit Survey, 1996
Calculations may be subject to rounding error

---

[5] See, for example, Michael Tanner, "Why Private Charity Works Better," *Baltimore Sun* (July 2, 1997), p. 17A.

new social needs, and government enabling nonprofit institutions to respond to such needs in a major way. This is a partnership that has long characterized government and nonprofit activity in America. It is a partnership that seems to have been especially prominent in the evolution of nonprofit action in Maryland, though the private partner has been far more limited than is the case elsewhere.

III. BRINGING IN VOLUNTEERS

Before resting content with this overview of the financing of nonprofit activity in Maryland, it is important to acknowledge that the discussion so far has focused exclusively on cash income. In fact, however, Maryland nonprofit organizations also receive considerable in-kind support, most notably in the form of volunteer time. To what extent does the picture of nonprofit revenue painted above change when the contribution of volunteers is factored into the equation?

To answer this question, we added the value of volunteer time to private giving and recalculated the shares of income accounted for by the different sources on this basis.[6]

Aggregate Picture

As Figure 4.5 indicates, the inclusion of volunteers does alter the aggregate picture of nonprofit finance, but not overwhelmingly. In particular:

- With volunteer time included, the share of private giving in total nonprofit revenue more than triples--from 4 percent to 15 percent.

- At the same time, private giving remains the third most important source of income for Maryland nonprofit agencies, and a distant third at that. Thus, compared to the 15 percent of nonprofit income that comes from all sources of private giving, including volunteerism, 39 percent comes from government and 46 percent from earned income.

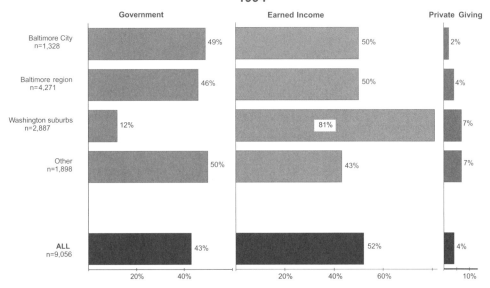

Figure 4.4 Sources of Income of Maryland Nonprofit Agencies, By Region, 1994

Source: MANO/JHU-IPS Maryland Nonprofit Survey, 1996

[6]For this analysis, the value of volunteer time was assumed to be equal to the average wage in the service sector.

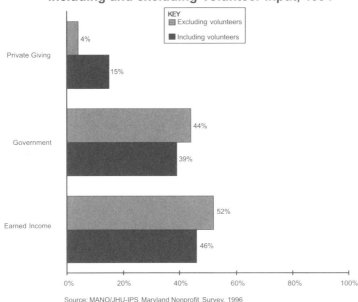

Figure 4.5 Sources of Income of Maryland Nonprofit Agencies, including and excluding Volunteer Input, 1994

Source: MANO/JHU-IPS Maryland Nonprofit Survey, 1996

Variations by Type of Agency

The inclusion of volunteer time does alter the revenue picture in a number of fields rather substantially, however. In particular:

- With volunteer time included, private giving becomes the dominant source of funding among two additional types of agencies--*social service* and *multiservice*. These agencies evidently absorb significant portions of the volunteer effort in the state.

- The inclusion of volunteer inputs also substantially boosts the role that private giving plays in the fields of *community development* and *culture and recreation*. In each of these, private giving accounts for a quarter or more of total income with volunteer contributions included.

- More generally, with hospitals and education institutions excluded, private giving emerges as the second largest source of nonprofit income when the value of volunteer time is included--ahead of government but behind earned income. Thus, private giving so defined accounts for 31 percent of the income compared to 42 percent for earned income and 27 percent for government.

CONCLUSIONS

The revenue structure of the Maryland nonprofit sector thus deviates markedly from popular conceptions, and does so even more than that elsewhere. Private giving, far from providing the bulk of the income, provides less than 5 percent, well below the national average. Although such giving is more important for the numerous small agencies than this figure would suggest, and plays a larger role once the value of volunteer effort is considered, the fact remains that the substantial growth that has been evident in the state's nonprofit sector would not have been possible had private giving been the only source of support. Rather, the real engine of nonprofit growth in this state has been the expansion of governmental support for many of the activities in which nonprofit organizations have wanted to engage. The growth of govern-

Figure 4.6 Sources of Income of Maryland Nonprofit Agencies, with Volunteers, by Type of Agency, 1994

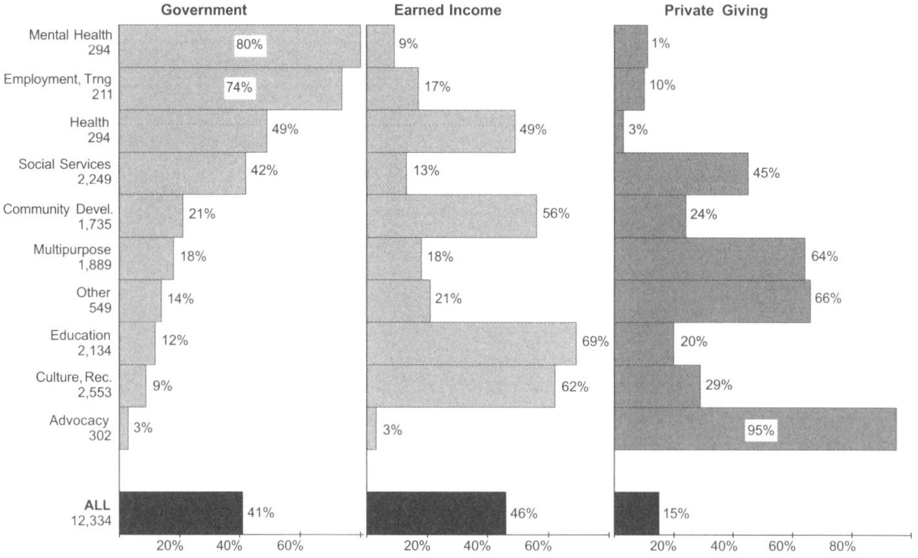

Type of Agency	Government	Earned Income	Private Giving
Mental Health (294)	80%	9%	1%
Employment, Trng (211)	74%	17%	10%
Health (294)	49%	49%	3%
Social Services (2,249)	42%	13%	45%
Community Devel. (1,735)	21%	56%	24%
Multipurpose (1,889)	18%	18%	64%
Other (549)	14%	21%	66%
Education (2,134)	12%	69%	20%
Culture, Rec. (2,553)	9%	62%	29%
Advocacy (302)	3%	3%	95%
ALL (12,334)	41%	46%	15%

Source: MANO/JHU-IPS Maryland Nonprofit Survey, 1996

ment has therefore hardly displaced the nonprofit sector, as many have alleged. Rather, it has empowered this sector and enabled it to broaden and deepen its reach. Most significantly, perhaps, it has allowed nonprofit agencies to increase their service to the poor. One of the more striking conclusions to emerge from this analysis, in fact, is that service to the poor has depended more on the availability of government support than on any other source.

While private giving has played a much smaller role than popularly assumed, however, it has nevertheless performed a very useful function in the nonprofit world. Private giving has functioned as a "venture capitalist" of the nonprofit sector, giving new organizations a start and sustaining them through a start-up phase. Even here, however, it has shared this function with government and encountered limits in promoting the take-off that is often needed.

All of this emphasizes the significance for Maryland nonprofit organizations of the debates now under way in our nation over the appropriate levels of government human service funding. What the Maryland experience demonstrates more clearly than that of most other states is the inappropriateness of framing this debate as a choice *between* government and nonprofit action. In Maryland, the choice has been between government *and* nonprofit action on one side, and no action at all on the other.

Beyond this, however, the data presented here also suggest quite strongly the need to buttress the existing levels of private charitable support in this state--not as a substitute for government, but as a complement to it. While nonprofit organizations can certainly grow on the basis of government and earned income, they can find it difficult to maintain their independence if these are the only sources of support. What is more, with government support in jeopardy, it seems likely that the state's nonprofit organizations will drift more fully into an essentially commercial mode of operation in the absence of significantly expanded charitable support. To understand this point, however, it is necessary to look in more detail at some of the challenges the Maryland nonprofit sector is facing, which is the focus of the next chapter.

CHAPTER FIVE
CHALLENGES AND RESPONSE

The funding structure of the Maryland nonprofit sector depicted in the previous chapter did not simply take shape in a vacuum, of course. Rather, it evolved over time in response to a variety of pressures and developments.

These pressures continue to affect the evolution of the state's nonprofit sector. Indeed, like their counterparts elsewhere, Maryland nonprofits have confronted enormous challenges in recent years. The purpose of this chapter is to look a bit more closely at these challenges and then to examine how Maryland nonprofits have responded to them. Given the limitations of our data, we can generally do this only indirectly, through the perceptions of agency managers. Because of this, it is often not possible to look back more than five, and often more than two, years. Despite these limitations, however, it is possible to develop a reasonable picture of the context within which Maryland nonprofits have been operating in recent years and of the ways they have sought to respond.

The general picture that emerges from such an analysis is of a set of organizations struggling to preserve their missions in the face of significant fiscal and other pressures, while slipping into increased commercial operations in self-defense and hoping for rescue from private, charitable support.

I. THE CHALLENGES

The past decade and a half has been a time of testing for nonprofit organizations in the United States.[1] After two decades of expansion in the 1960s and 1970s fueled in important part by growth in government support, nonprofit organizations experienced a significant fiscal shock as a result of budget cuts in the early 1980s. These cuts were partly reversed later in the decade, but a renewed period of fiscal stringency began with the election of the "Contract with America" Congress in 1994. Meanwhile, global economic shifts that undermined the position of unskilled workers in American cities and changes in lifestyles and demographics boosted the need for the services that many nonprofits provide. At the same time, increased competition from for-profit providers narrowed the maneuvering room that such organizations enjoyed. On top of this, finally, a series of highly publicized scandals seemed to sour public attitudes towards nonprofit institutions, threatening the public confidence on which they ultimately rely.

To what extent are these pressures evident among the Maryland nonprofit agencies we surveyed? What effect have they had on agency operations and agency ability to carry out their missions?

To answer these questions, we asked Maryland nonprofit managers a variety of questions about the challenges they have been facing in recent years. The results suggest a set of institutions under considerable strain in a variety of respects.

Increased Demand for Services

In the first place, the early 1990s was a period of expanding demand for the services

---

[1] For a summary of these challenges, see Lester M. Salamon, *Holding the Center: America's Nonprofit Sector at a Crossroads* (New York: The Nathan Cummings Foundation, 1997).

that Maryland nonprofit agencies provide. Nearly two-thirds of the agencies (64 percent) reported noticeable increases in the demand for their services during the previous two years, and for nearly 40 percent the increases were substantial (i.e. 10 percent or more).

As shown in Figure 5.1, increased demand did not vary much by the size of the organization, but it was far more prevalent among agencies in some fields than others. In particular, social service and health providers were especially likely to register substantial increases in demand. Among the former, in fact, nearly three out of five reported a substantial increase in demand over the two years prior to our survey.

Constrained Income Growth

In the face of this growth in demand, nonprofit agencies in Maryland reported only modest growth in income. Although 45 percent of Maryland agencies reported increases in income between 1989 and 1994, for only 14 percent did the reported increase exceed the rate of inflation.[2] This means that 86 percent of Maryland's nonprofit agencies were not able to boost their incomes enough to keep pace with inflation. Put somewhat differently, the real incomes of five out of six Maryland agencies actually declined between 1989 and 1994.

Variations among Types of Agencies. These declines were not spread evenly among types of agencies, however. To the contrary, as Figure 5.2 shows:

• Large agencies performed much better than small or medium-sized agencies. Thus, compared to the 14 percent of all agencies that experienced real income growth, the share of large agencies that managed to achieve this level of growth was 34 percent. The overall income of the sector may have grown, therefore, even though most agencies failed to keep pace with inflation.

• In the second place, education and research organizations reported real income growth somewhat more commonly than other types of agencies. Thus, over 20 percent of the education organizations reported real income growth, compared to only 14 percent of the human service and other organizations and 6 percent of the culture and recreation organizations.

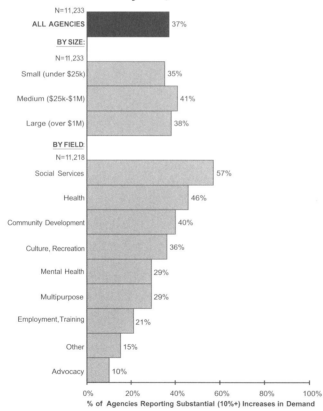

Figure 5.1 Changes in Demand for Nonprofit Services in Maryland, 1994-96

Source: MANO/JHU-IPS Maryland Nonprofit Survey, 1996

---

[2]Inflation in the service sector grew by 23 percent between 1989 and 1994. To exceed inflation, therefore, agencies had to experience income growth in excess of 20 percent during this period, which only 14 percent of the agencies did. The inflation estimate is based on the Chain-Type Price Index for Service Expenditures as reported in: *Economic Report of the President* (Washington: U.S. Government Printing Office, 1997), p. 304.

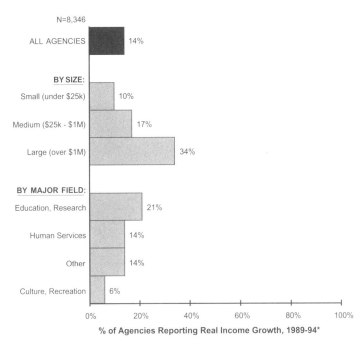

Figure 5.2 Share of Maryland Nonprofit Agencies with Real Income Growth, 1989-94, by Size and Type of Agency

* Represents agencies reporting increase in income of 20% or more, enough to exceed 23% inflation rate during this period.

Source: MANO/JHU-IPS Maryland Nonprofit Survey, 1996

Variations by Income Sources. What lies behind this generally disappointing record of recent income growth for the overwhelming majority of Maryland nonprofit organizations is a fairly tepid pattern of growth for most of the major income sources. Thus, as Figure 5.3 shows:

- Fewer than 10 percent of the agencies reported five-year increases in income from any source that exceeded the rate of inflation. Thus only 7 percent reported such increases from service fees, only 7 percent from individual giving, and only 6 percent from government.

- For the other sources of income, the share of agencies that reported real growth between 1989 and 1994 was smaller still--6 percent for corporate support, 4 percent for foundation support, and 2 percent for United Way and other federated funders.

- Interestingly, a larger proportion of agencies registered substantial income increases from "special events" than from United Way, foundations, or other federated fundraising campaigns; or from related and unrelated businesses.

This overall pattern of income change did not apply across the board, however. Rather, the larger agencies seem to have performed better with respect to all the income sources than did the small or medium-sized agencies. More specifically, as Table 5.1 shows:

- Large agencies apparently had greater success in boosting government and fee income. Thirty and 23 percent, respectively, of all large agencies reported increases from these two sources that exceeded the rate of inflation. Since these are the two largest sources of nonprofit income, this finding makes clear again that the state's nonprofit sector may have grown considerably in aggregate scale during this period even though most agencies reported no, or limited, real growth.

- Above-average proportions of large agencies also boosted their real income from investments and special events. Thus, while 3 percent of all agencies reported substantial

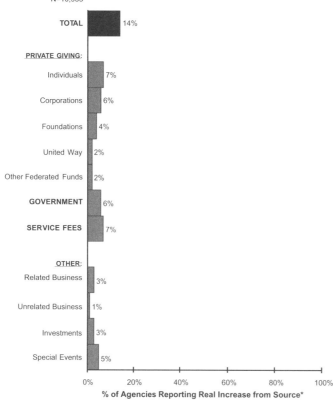

Figure 5.3 Share of Maryland Nonprofit Agencies with Real Income Growth, 1989-94, by Source
N=10,935

Source: MANO/JHU-IPS Maryland Nonprofit Survey, 1996
*Increases of 20% or more between 1989 and 1994

increases in related business and investment income, the proportion of large agencies that did so was 12 percent and 16 percent, respectively.

• Fewer differences are evident among agencies defined in terms of their fields of service. Generally speaking, education organizations were somewhat more successful in generating individual charitable contributions, fee income, and government support; and human service organizations did better with government and fee income.

In short, in the face of increased demands for their services, most Maryland nonprofit organizations had a difficult time keeping pace with inflation, let alone expanding their activities. At the same

Table 5.1
Recent Changes in Operating Revenues of Maryland Nonprofit Agencies, 1989-94, by Size of Agency and Source of Income

	% of Agencies Reporting Large (Over 20%) Increases			
	All Agencies	Small Agencies	Medium Agencies	Large Agencies
Total Income	14%	10%	17%	34%
Private Giving				
Individuals	7	5	10	13
Corporations	6	4	8	11
Foundations	4	-	10	12
United Way	2	-	5	4
Other Federated Funds	2	-	3	9
Government	6	1	12	30
Fees, Charges, Other	7	3	13	23
Related business income	3	-	6	12
Unrelated business income	1	-	2	2
Investment income	3	1	5	16
Special events	5	4	6	12

Source: MANO/JHU-IPS Maryland Nonprofit Survey, 1996

time, the larger agencies were considerably more successful at this than the small and medium-sized agencies. Since these large agencies command the largest proportion of the sector's resources, the aggregate income of the sector may have kept pace with inflation even though only a minority of the agencies participated in the growth. The general picture, however, is one of considerable constraint.

Staffing Challenges

The fiscal restraint facing most Maryland nonprofit organizations naturally had implications for nonprofit staffing. Interestingly, only 30 percent of the agencies reported increasing their numbers of full-time paid staff over the previous two years, though among the larger agencies this figure reached almost 50 percent. More generally, agencies reported a considerable range of staffing challenges that were very likely related to the income constraint many of them faced. For the most part, these challenges confronted all types of agencies about equally, though in a few cases the impact varied by agency size. In particular, as portrayed in Figure 5.4:

- **Problems with staff recruitment and retention.** Nearly half of all Maryland nonprofit organizations reported problems recruiting able paid staff, and for about a quarter of them the problems were serious or very serious. Significantly, moreover, these problems were particularly severe among the larger agencies, 40 per-

cent of which reported "serious" or "very serious" problems *recruiting* professional staff and 30 percent of which reported "serious" or "very serious" problems *retaining* paid staff.

- **Pressure on staff salaries and benefits.** One important reason for these staffing problems is very likely the pressure that apparently exists on staff salaries. Three-fourths of Maryland agencies reported some problem in offering competitive salaries to their staff, and for about half of the agencies--including roughly proportional numbers in each size class--these problems were "serious" or "very serious." Indeed, over half of the agencies reported no change or actual declines in average salary levels over the previous two years.

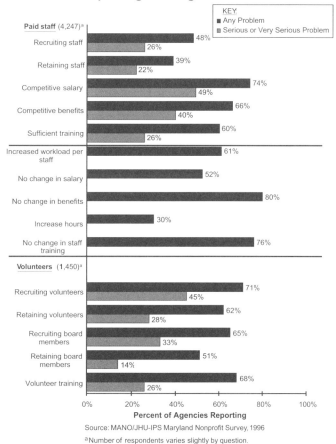

Figure 5.4 Share of Maryland Nonprofit Organizations Reporting Staffing Problems

Source: MANO/JHU-IPS Maryland Nonprofit Survey, 1996
[a] Number of respondents varies slightly by question.

Similar problems were reported, moreover, with respect to benefits. Two-thirds of the agencies reported problems maintaining competitive benefit packages for their employees, and for 40 percent these problems were "serious" or "very serious." Significantly, such benefit problems were especially prevalent among the small and medium-sized agencies, which lack the economies of scale often required to offer medical and other benefits at reasonable cost.

- **Increased workloads.** In addition to facing downward pressures on their salaries and benefits, nonprofit employees are also experiencing increases in their workloads. Thus, over 60 percent of the agencies reported increases in the workload per paid employee over the previous two years and 30 percent reported an increase in the typical hours of work per week. Nonprofit agencies thus seem to be demanding more of their employees while finding themselves unable to reward the increased work with higher pay.

Volunteer Challenges

Not only are Maryland nonprofit organizations experiencing problems recruiting and retaining *paid* staff, they are also experiencing problems with the volunteer side of their operations. Thus, as Figure 5.4 also shows:

- Over 70 percent of the agencies reported difficulties recruiting dependable, qualified volunteers, and 62 percent reported difficulties retaining such volunteers once recruited;

- Two-thirds of the agencies also reported problems recruiting board members, and one- third classified these problems as "serious" or "very serious." These problems do not seem to vary much by agency size, moreover.

- Finally, close to 70 percent of the agencies reported problems providing sufficient training for volunteers. This is significant because volunteers can typically not be used effectively without adequate training and preparation.

What these data make clear is that the volunteer "solution" is hardly a panacea for the staffing problems confronting Maryland nonprofit organizations. To the contrary, volunteerism itself faces important challenges in the state.

Public Attitudes

In addition to the fiscal pressures and resulting staffing difficulties they are confronting, Maryland nonprofit organizations are also facing important external threats at the present time. Perhaps the most serious of these is an apparent lack of public support. This phenomenon is not, of course, peculiar to Maryland. Recent national polls suggest that only a third of the American public feel "a great deal" or "quite a lot" of confidence in nonprofit organizations other than hospitals and higher education institutions. This is slightly above the share of the population that expresses similar degrees of confidence in the federal government or state and local government, but it is below the confidence levels enjoyed by the military and small business.[3] The reasons for this apparent lack of confidence are numerous, but the recent scandals involving pay and perquisites for highly visible nonprofit leaders and the general assault on the effectiveness of government social programs in which nonprofit organizations have been deeply involved have undoubtedly played a role.

> *The volunteer "solution" is hardly a panacea for the staffing problems confronting Maryland nonprofit organizations. To the contrary, volunteerism itself faces important challenges in the state.*

---

[3]Independent Sector, *Giving and Volunteering 1994* (Washington: Independent Sector, 1994), p.54.

While our survey does not allow us to determine how widespread such attitudes are within the Maryland public, or whether they have changed over time, it does permit us to determine the extent to which nonprofit executives believe they exist. In particular, as Figure 5.5 reports:

- Over half of the nonprofit executives we surveyed do report that they believe that "the public is becoming increasingly distrustful of nonprofit organizations."

- The executives generally do not believe this distrust is warranted, however. By overwhelming majorities, they are convinced that nonprofit services make a difference, that the need for nonprofit organizations is greater than it was five years ago, and that nonprofits are effectively demonstrating the value of what they are doing. What is more, only a minority (32 percent) of the agency heads buys into the argument that some conservative critics have recently advanced to the effect that nonprofit status should be reserved only for agencies primarily serving the needy.[4]

- The principal explanation of declining public support advanced by Maryland nonprofit leaders, rather, has to do with "compassion fatigue"--i.e.excessive solicitation of the public--and the scandals of the past few years. Thus 48 percent of the agencies indicated their belief that there are "too many" nonprofits soliciting charitable contributions from the public at the present time, and 44 percent endorsed stronger enforcement of state charitable solicitation laws. While only about 40 percent of the agencies indicated that "on average, nonprofit executives are paid too much," close to 60 percent agreed that there is a need for "an effective system of self-regulation" within the nonprofit sector and endorsed the adoption of a stronger nonprofit codes of conduct. In addition, only about

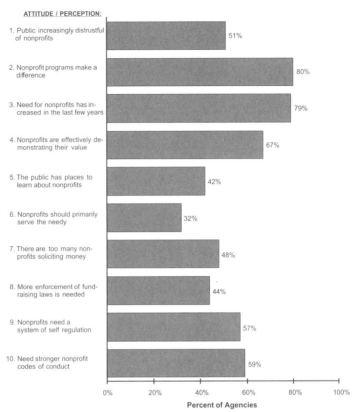

Figure 5.5 Perceptions of Public Attitudes among Maryland Nonprofits

ATTITUDE / PERCEPTION:
1. Public increasingly distrustful of nonprofits — 51%
2. Nonprofit programs make a difference — 80%
3. Need for nonprofits has increased in the last few years — 79%
4. Nonprofits are effectively demonstrating their value — 67%
5. The public has places to learn about nonprofits — 42%
6. Nonprofits should primarily serve the needy — 32%
7. There are too many nonprofits soliciting money — 48%
8. More enforcement of fundraising laws is needed — 44%
9. Nonprofits need a system of self regulation — 57%
10. Need stronger nonprofit codes of conduct — 59%

Source: MANO/JHU-IPS Maryland Nonprofit Survey, 1996

> *Over half of the nonprofit executives we surveyed do report that they believe that "the public is becoming increasingly distrustful of nonprofit organizations."*

---

[4] See for example: Alan Reynolds, *Death, Taxes and Charity* (Washington, D.C.: Philanthropy Roundtable, 1997). This idea was partially enshrined in a "tax credit" proposal originally championed by Senator Dan Coats of Indiana and later adopted by Republican Presidential candidate Robert Dole in 1996.

40% of the agencies believed that the public has adequate places to turn for reliable information about nonprofit organizations.

In short, the Maryland nonprofit leaders we surveyed remain convinced of the worth of the work they are doing and attribute recent declines in public confidence to ethical lapses on the part of a handful of nonprofit leaders rather than a more basic misunderstanding or lack of understanding on the part of the public about the nature of nonprofit organizations and the role that they play. This is a hopeful perspective. Whether it is a realistic one or not is harder to say.

For-Profit Competition

A fourth major challenge confronting the nonprofit sector nationally is the growth of for-profit competition in traditional areas of nonprofit operation.[5] As nonprofit organizations have turned increasingly to fees and charges to offset reductions in government support, they have demonstrated the existence of quasi-commercial markets in many of the fields in which they operate—day care, home health, family counseling, drug abuse treatment, health care, employment and training, and others. This has, in turn, attracted for-profit firms into these markets. Unlike the nonprofit providers, moreover, the for-profits can more easily limit their focus to the most profitable market segments and clients. Not only can they thereby undercut the nonprofits on price, but also they can siphon off the most profitable clients and limit the ability of nonprofit organizations to generate revenues they need to cross-subsidize services for those less able to pay. This result can be to close off an important source of fiscal stability for the nonprofit organizations.

As we saw in Chapter Two above, this trend seems to be somewhat less evident in Maryland than it is in the nation as a whole. Perhaps reflecting this, only 26 percent of the Maryland nonprofit organizations in our survey indicated that they were concerned about the adverse impact of competition from for-profit providers.

This general conclusion may be a bit misleading, however, since there is clear evidence of concern about for-profit competition in some spheres. In particular, as reflected in Table 5.2, while only 22 percent of the *small* agencies perceive a threat from for-profit firms, over half of the *larger* organizations do. This is significant because the larger firms

Close to 60 percent agreed that there is a need for "an effective system of self-regulation" within the nonprofit sector and endorsed the adoption of stronger nonprofit codes of conduct.

Table 5.2
Concerns about For-Profit Competition Among Maryland Nonprofit Agencies, by Size of Agency
(n=11,309)

Size Class	For-Profit Competition will Adversely Affect our Operations	
	Agree	Disagree
Small	22%	65%
Medium	29	56
Large	54	34
All	**26%**	**60%**

Source: MANO/JHU-IPS Maryland Nonprofit Survey, 1996

---

[5]Salamon, *Holding the Center*, pp. 29-37.

represent the bulk of all the nonprofit activity. Evidently, competition from for-profit firms is more serious than the overall survey results might suggest.

Conclusion

In short, Maryland nonprofit agencies are confronting significant challenges at the present time. In the face of escalating demand for their services, nonprofit revenues have stalled. Only one in five agencies has thus been able to keep revenues in line with inflation. This has, in turn, created significant staffing pressures--increased staff workload, constraints on staff salaries and benefits, and a resulting greater difficulty recruiting and retaining able staff. All of this is taking place, moreover, at a time of strained public support and growing concerns about competition from for-profit providers.

II. NONPROFIT RESPONSES

How have Maryland nonprofits responded to these challenges? To what extent is there evidence that the state's nonprofit agencies are taking reasonable corrective action to offset the combined fiscal, staffing, attitudinal, and competitive challenges that they face? And what impact is this having on those the agencies are supposed to serve?

The discussion above has already hinted at part of the answer to these questions. As we saw, one of the major responses nonprofit agencies seem to have made to the pressures they are under is to shift these pressures, at least in part, to their own personnel--by increasing staff workloads and work hours. But is this the extent of the reaction? What other strategies are agencies pursuing, and with what result?

Protecting the Client Base

A useful starting point for this discussion is the client focus of the agencies. Nonprofit organizations have significant discretion in determining whom they will serve. The more reliance is placed on fees and service charges, for example, the less an agency can respond to the needs of the poor. By increasing existing fees or charging fees where none existed, agencies can subtly alter their client focus or generate greater income from existing clients, at least some of whom are in a position to pay. Similarly, agencies can expand their operations in areas where paying clientele predominate in order to buttress their financial base, even if this involves reducing their focus on the poor. Since these changes can often be made without major alterations of existing operations, they are attractive strategies for hard-pressed executives.

To what extent do we find evidence that Maryland nonprofits have been moving in these directions in recent years?

The answer that our survey produces to this question is somewhat ambiguous but generally encouraging:

> *While only 22 percent of the small agencies perceive a threat from for-profit firms, over half of the larger organizations do.*

> *A quarter of the agencies reported that funding pressures were pushing them toward "becoming less responsive to client needs and more responsive to market forces" and among the large agencies this reached fully half of the agencies.*

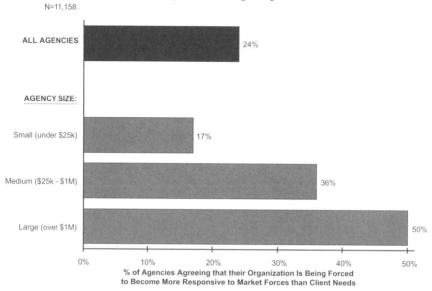

Figure 5.6 Share of Agencies Reporting Increased Market Pressures, by Size of Agency

- In the first place, as Figure 5.6 shows, a quarter of the agencies reported that funding pressures were pushing them toward "becoming less responsive to client needs and more responsive to market forces." Among the large agencies, moreover, where reliance on fee income is most in evidence, fully *half* concurred with this statement.

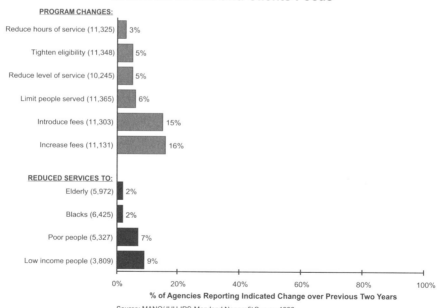

Figure 5.7 Share of Maryland Nonprofit Agencies Reporting Recent Changes in Service Levels and Clients Focus

- Presumably reflecting these pressures, some 15-16 percent of agencies introduced new client fees or increased those already in operation, as reflected in Figure 5.7. Among the large agencies, moreover, these practices were even more widespread. Thus, 35 percent of the larger organizations reported introducing new client fees during the past two years, and 61 percent reported increasing the fees they already charged.

- At the same time, however, as Figure 5.7 also shows, agencies appear to be resisting more direct changes that would restrict client access. Thus, only 5 percent tightened their eligibility requirements, only 6 percent reduced the number of services they provide or the number of people served, and only 3 percent reduced their hours of service.

- Perhaps because of this, there is little evidence that the changes introduced in recent years have significantly altered the client focus of the agencies. Thus, only 2 percent of the agencies reported decreasing their service to the elderly, only 2 percent reported decreasing service to Blacks, only 7 percent reported decreasing service to the poor, and only 9 percent reported decreasing their service to persons whose income is within 150 percent of the poverty line. These percentages do not vary, moreover, by the size of the agency. In other words, the large agencies, which were much more likely to introduce or raise fees, were not more likely to reduce their service to the poor. To the contrary, if anything, they were more likely to increase such services. *Evidently, the agencies turning more heavily to fees and charges are doing so mostly with respect to their less needy clientele, and are doing so at least in part in order to be able to sustain their focus on those in greatest need in the face of budget and other pressures pushing them in the opposite direction.* It is for this reason, however, that competition from for-profit providers poses such a serious challenge to the nonprofit sector: by siphoning off more profitable clients, for-profit firms may deny nonprofits income they need to provide free or reduced-cost services to those unable to pay.

> *There is little evidence that the changes introduced in recent years have significantly altered the client focus on Maryland's agencies.*

Management Changes

Whether Maryland nonprofits are able to sustain their present degree of attention to the poor into the future will depend, however, on what other changes they make in the face of the challenges they are confronting. One of these other types of changes are managerial and organizational in character. Generally speaking, the larger agencies are more active than the smaller ones in making such changes, but considerable proportions of agencies of all sizes seem to be making efforts along these lines, and even larger proportions are planning to do so in the immediate future. Thus, as shown in Figure 5.8:

- Almost half of all agencies are making more extensive use of volunteers in their basic program operations;

- About 45 percent of all agencies, and 90 percent of the large agencies, have recently instituted new management practices to increase their efficiency or improve agency functions, with over half the agencies planning to do so in the near future;

- Other widespread management changes adopted over the past two years involve increasing *user involvement* (28 percent of agencies), *sharing resources with other agencies* (27 percent), *increased staff training* (27 per-

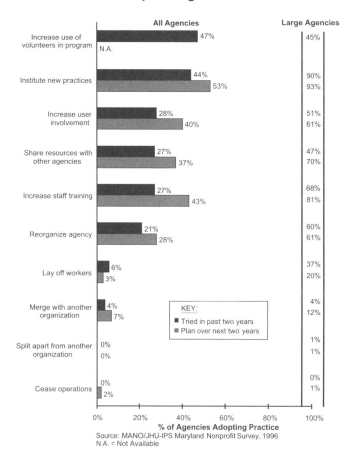

Figure 5.8 New Management Practices of Maryland Nonprofit Agencies

Fundraising Changes

Perhaps the major strategy Maryland nonprofit organizations are relying on to withstand the pressures they are under, however, is to change their fundraising practices. Nearly two-thirds of Maryland's nonprofit organizations report that they tried at least one new fundraising approach in the previous two years, and three-fourths report that they are planning additional changes over the next two years. The adoption of new fundraising practices is particularly widespread among the larger agencies, but large numbers of small and medium-sized agencies are also pursuing new sources of funds. In fact, the average Maryland agency had tried at least 2 new fundraising approaches over the past two years, and expected to try 3 over the next two years.

cent), and *reorganizing agency functions* (21 percent).

- Far less common are staff reductions and mergers with other agencies, though a surprising 7 percent of all agencies indicated their intention to take this latter step over the next two years.

While hardly fundamentally restructuring their operations, Maryland nonprofit organizations do seem to be experimenting with management changes that can improve their efficiency and make them more responsive to user concerns.

Types of new fundraising practices. As reflected in Figure 5.9 below:

- The most commonly tried new fundraising practice over the past two years was *applying to new corporate and foundation donors*. A third of all agencies utilized this approach in the past two years and half expect to make a try over the next two years. This approach was particularly popular among the largest agencies, 75 percent of which reported approaching new corporate or foundation funders in the previous two years, and 94 percent of which plan to approach such funders over the next two years.

- *Special events fundraising, product sales, and new government programs* were the next most commonly used new fundraising techniques over the past two years. About a quarter of the agencies reported trying these techniques over the previous two years and between 35 percent and 50 percent plan to turn to them over the next two years. Large agencies are no more likely than small and mid-sized agencies to try the first two of these approaches, but they are much more likely to pursue new government programs. In fact, half of the large agencies reported applying to such programs over the previous two years, and 70 percent expect to do so in the next two years.

- As noted earlier, a relatively small proportion (16 percent) of the agencies increased their *fees* over the previous two years and only a quarter plan to do so in the next two. Among the large agencies, however, the proportion turning to this source was over 60 percent.

- Aside from increased sales of products, relatively small proportions of agencies reported turning to explicitly commercial activities to meet the fiscal pressures they are facing. Thus only 5 percent reported establishing a *joint venture with a for-profit company,* and only 4 percent reported setting up a *for-profit subsidiary.* While these techniques are generating considerable interest in press accounts, they are thus attracting far less engagement on the part of actual nonprofit organizations.

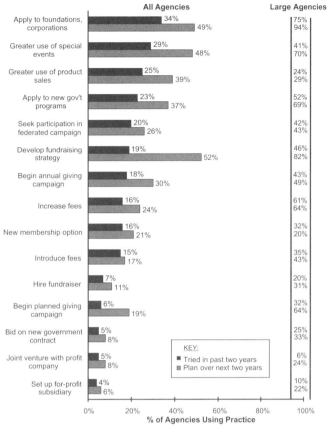

Figure 5.9 New Fundraising Practices of Maryland Nonprofit Agencies

Source: MANO/JHU-IPS Maryland Nonprofit Survey, 1996

Anticipated Funding Trends. Maryland nonprofit organizations are rather optimistic about the likelihood that these new fundraising strategies will yield results. Thus, as shown in Table 5.3, over half (55 percent) expect to boost their revenues by at least 10 percent over the next two years, and there are no significant differences between large and small agencies in this respect.

These expectations prevail, moreover, despite the fact that only a small proportion of agencies (18 percent) expect substantial increases in government support over the next two years, and this is now true not only for the

Table 5.3
Anticipated Changes in Revenues of Maryland Nonprofit Organizations, by Source, 1996-98
n = 8,272

Percent of Agencies Anticipating Moderate (10-20%) or Large (over 20%) Income Increases Over the Next Two Years

Source	All Agencies	Small Agencies	Medium Agencies	Large Agencies
Total Income	55%	53%	61%	55%
Private Giving				
Individuals	47	47	44	57
Corporations	28	24	34	43
Foundations	21	15	29	42
United Way	9	6	15	11
Other Federated Funds	8	4	14	16
Government	18	16	21	24
Fees, Charges	25	15	36	45
Other				
Related business	10	6	16	23
Unrelated business	2	2	2	7
Investment income	11	2	24	39
Special events	34	33	36	35

Source: MANO/JHU-IPS Maryland Nonprofit Survey, 1996

smaller agencies but for the large ones as well. Reflecting this, about half of the agencies reported that they expected that the effort to balance the federal budget would adversely affect their agency. Among the large agencies, moreover, this figure reached 84 percent. In other words, Maryland nonprofits are anticipating another period of significant fiscal stringency with respect to revenues from government programs, one of the principal sources of local nonprofit support. How, then, do they propose to generate the anticipated growth?

The answer, as reflected in Table 5.3, is far from clear:

- **Expectations for individual support.** Agencies are pinning particular hopes on *individual giving*. Just under half of all agencies and close to 60 percent of the large agencies anticipate substantial increases from this source.

- **Expected growth in special events.** A second major source of hope for Maryland nonprofit agencies is special events fundraising. A third of all agencies are looking to this source to generate significant income growth, and this is as true of small agencies as large ones.

- **Corporate and foundation support.** In addition to individual giving and special events, significant proportions of Maryland's nonprofit agencies are anticipating moderate to large increases in *corporate and foundation support*. These expectations are particularly prevalent among the larger agencies, over 40 percent of which expect moderate or substantial increases from these sources.

- **Fee income.** Finally, close to half of the larger agencies, though much smaller proportions of the small ones, expect substantial further increases in their *fee income*.

Given the patterns of fundraising described in Chapter Four above, there is reason to be highly skeptical about these expectations. As we have seen, all of private giving combined accounts for only 4 percent of nonprofit income in Maryland, compared to the 40 percent of income that comes from government. To offset a 10 percent decline in government support, therefore, private giving would have to increase by 100 percent. Similarly, special events currently contribute only 3 percent of total nonprofit income. The two major sources to which Maryland nonprofit organizations are looking for increased funding thus seem highly unlikely to yield the returns expected. This means that the most likely outcome of the fiscal stringency many agencies are anticipating is further reliance on fees and service charges. Over time, however, it seems likely that increased reliance on this source will erode the ability of agencies to maintain even their existing levels of attention to the poor.

CONCLUSION

Maryland nonprofit agencies thus find themselves buffeted by a variety of challenges at the present time. In the face of escalating demand, government support has stopped growing and in many cases begun to decline. At the same time, the larger agencies are facing expanded competitive threats. To cope with these changes, the agencies have launched concerted efforts to locate new sources of funds, to streamline their operations, and to mobilize additional volunteer help. To date, these efforts appear to have enabled these organizations to avoid making painful reductions in their staffs or in the clientele that they serve. Whether they will be able to so into the future, however, seems far more problematic.

CHAPTER SIX
ETHICS AND ACCOUNTABILITY[1]

One of the crucial determinants of the ability of Maryland nonprofit organizations to remain viable into the future is the extent to which they retain public confidence and support. In recent years, however, the public image of the nonprofit sector has been threatened at the national level by front page headlines featuring stories of ethical lapses, indiscretions, mismanagement, and outright fraud in the operation of individual organizations. While it is possible to view these events as isolated instances of misconduct perpetrated by a few bad actors, there is clear evidence that they have affected perceptions about the nonprofit sector in Maryland as well. As we have seen, half of the agency directors we surveyed indicated their belief that the public is becoming increasingly distrustful of nonprofit organizations.

The purpose of this chapter is to assess the extent to which such public distrust is warranted in the case of Maryland nonprofit agencies. More specifically, it explores the internal management practices of Maryland's nonprofit agencies to determine how well they adhere to accepted standards in the field and to the expectations of nonprofit managers themselves.

> *Nonprofit organizations generally lack the built-in performance measures that operate in the business world. Therefore, other devices have to be used.*

I. STANDARDS OF ETHICAL MANAGEMENT

To put the discussion here into context, it is useful to acknowledge the special ethical problem that nonprofit management entails, and the devices that have been developed to deal with it.

In the for-profit business world, the goal of profit maximization and the presence on corporate boards of persons who have a financial interest in the success of the organization provide a built-in system of accountability. To buttress this, moreover, legal protections have been created in the form of an elaborate system of securities regulation to ensure that corporations report regularly to the public, and that their financial reports adhere to generally accepted accounting practices.

For the most part, these built-in mechanisms are absent in the nonprofit world. Nonprofit organizations are significantly insulated from a direct market test. Although they have boards of directors, these boards do not have the kind of personal stake in the performance of the organization that corporate boards have. Finally, nothing akin to the Securities and Exchange Commission exists to protect the public stake in the performance of nonprofit organizations. Although states have charity supervision offices that provide some

---

[1]This chapter was co-authored with Peter Berns.

protection against unscrupulous or misleading fundraising campaigns, the legal authority of such offices is generally limited and consists mainly of registration and information requirements. Similarly, the federal Internal Revenue Service has limited authority to control the behavior of charitable institutions, and virtually no way to ensure the effectiveness of nonprofit activities.

To the extent nonprofit institutions are to be held accountable for their behavior, therefore, other approaches are needed. In particular, ethical behavior in the nonprofit sector is thought to be ensured through four principal devices:

- first, by vesting responsibility in a volunteer board of directors;

- second, by requiring that this board subordinate the self-interest of its members to the public interest that the organization is supposed to serve;

- third, by stipulating that the organization clarify its mission and gear all behavior to it; and

- finally, by requiring that the organization's stakeholders have access to the information they need to judge the organization's performance.

In the balance of this chapter we examine the extent to which Maryland nonprofit leaders subscribe to this basic set of management principles and then assess the extent to which the organizations they direct actually operate in conformance with them. What emerges most clearly from this analysis is the conclusion that concerns about the ethical performance of Maryland nonprofits may be overstated. At the same time, it is clear that actual management practice within the sector lags behind the managers' own vision of how nonprofits should act.

II. THE BOARD OF DIRECTORS: THE PIVOT OF NONPROFIT MANAGEMENT

In the absence of stockholders, control of a nonprofit corporation under Maryland law is vested in the "members" of the corporation. Most often, the sole members of a nonprofit corporation are the organization's board of directors, though in some instances an organization may have a broader class of "members" who in turn elect representatives to serve on the organization's board. Under either structure, the power and responsibility for governance of the organization rests in the hands of the board of directors.

Since the board is central to the governance of a nonprofit organization, it follows that how well boards function fundamentally affects ethics and accountability in the nonprofit sector. To the extent boards are functioning well, they provide direction and oversight which helps to prevent ethical lapses and to assure that resources are being used in furtherance of organizational missions. To the extent they are functioning poorly, boards may allow problems to develop and persist, including situations where a board member may misuse his/her position for personal benefit.

Criteria and Expectations for Effective Board Operations

But what are the features of a well-functioning board? Broadly speaking, they fall under several headings.

- First, the *composition* of the board must be appropriate. This means, at a minimum, that the Board should have a sufficient number of members so that control is not limited to a few individuals, nor is any board member able to dominate discussion or debate. Beyond this, since nonprofit board members serve in a fiduciary capacity, they should be free of any financial interests which could affect the loyalty that is owed to the nonprofit corporation. They can therefore not utilize their position to advance their personal interests. Thus, federal tax laws prohibit a nonprofit corporation from distributing its profits, if any, to private individuals. Nor may an organization operate in a manner which confers benefits on private individuals or allows private inurement for insiders. Where potential conflicts may exist, procedures have to be established to handle them.

- Second, boards should meet with sufficient frequency to govern the affairs of the corporation effectively.

- Finally, and perhaps most importantly, board members should be meaningfully involved in governing and setting policy for the organization, and in evaluating results, though without becoming immersed in the minutiae of day-to-day operations.

Maryland nonprofit managers generally endorse these basic criteria of effective board operations. Thus, as reflected in Figure 6.1:

- Three quarters (74 percent) of the respondents to our survey agreed that boards should have at least 5 members, though the extent of agreement was higher among medium-sized (85 percent) and large (88 percent) organizations than among small organizations (68 percent).

- Three quarters of respondents also agreed that boards should meet at least four times each year, and more than three quarters (79 percent) of respondents felt that there should be mandatory, minimum attendance requirements for board members.

- There was also widespread (81 percent) agreement among respondents that nonprofit board members should serve without compensation other than reimbursement for out-of-pocket expenses.

- Finally, there was widespread agreement that boards should play an active part in the operation of agencies. Indeed, no more than 7 percent of all the managers we surveyed felt

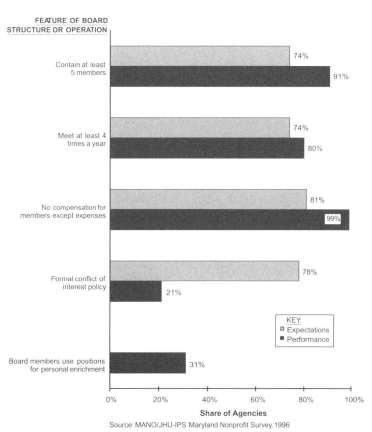

Figure 6.1 Structure and Operations of Maryland Nonprofit Boards: Expectations vs. Performance

Source: MANO/JHU-IPS Maryland Nonprofit Survey, 1996

their boards should be less involved than they were in any of 15 aspects of organizational management we explored. To the contrary, most favored greater board involvement in most areas.

How well do Maryland nonprofit boards live up to these expectations? By and large, they do quite well, though some problem areas are also apparent.

Performance I: Board Structure

In the first place, Maryland nonprofit boards appear to meet the basic structural and operational characteristics thought to be most conducive to effective board functioning. Thus, as Figure 6.1 shows:

- Nine out of ten boards contain 5 or more members. In fact, the average number of board members per organization is 12, with smaller organizations averaging slightly fewer (9 members) and large organizations considerably more (24 members).

- Similarly, Maryland nonprofit boards meet with sufficient regularity. In fact, 80 percent of the organizations report that their boards meet at least four times a year.

Performance II: Organizational Interests vs. Private Interests

A somewhat more complex picture surfaces when it comes to the management of the fiduciary responsibilities of Maryland nonprofit boards. Thus, as Figure 6.1 shows:

- Virtually no nonprofit board members are compensated for their service. Nearly one hundred percent (99.5 percent) of respondents reported that their board members are not compensated other than reimbursement for expenses, and even this is provided by fewer than twenty percent (19.3 percent) of the organizations.

Virtually no nonprofit board members are compensated for their service.

- At the same time, however, a significant minority of agencies (31 percent) reported that their board members use their positions for "personal enrichment," though it is at least possible that respondents had in mind aspects of "personal enrichment" that do not involve personal financial gain (e.g. increased knowledge, broadening of horizons, new acquaintances).

- Beyond this, 9 percent of the respondents reported that their organizations had purchased goods or services from a board member or a board member's immediate family in the past year. Among medium-sized organizations, moreover, this reached 11.4 percent, and among large agencies, 31 percent.

- Given this level of potential financial conflict, the fact that nearly 80 percent of the agencies favor formal conflict of interest policies is comforting. However, as Figure 6.1 reports, only 21 percent of the agencies report having such policies formally in place.

- Only among larger nonprofits, in fact, does a majority (62 percent) have a written conflict of interest policy. For medium and smaller nonprofits, written conflict of interest policies are the exception rather than the rule (10 percent and 33 percent, respectively).

Only one in five Maryland nonprofit agencies has a written conflict of interest policy in place.

These data thus suggest an important omission in the management practices particularly of small and medium-sized Maryland nonprofit agencies, and one that could affect the ethical performance of the sector. Here is therefore an opportunity for improvement in agency operations.

Performance III: Management Responsibilities

Beyond their narrow responsibilities to avoid conflicts between their personal interests and the interests of the organizations on whose boards they sit, nonprofit boards have a broader responsibility to take their management role seriously. This means that they must engage actively in the setting of organizational policies and the monitoring of organizational performance. How well do Maryland nonprofit boards measure up against this standard?

Figure 6.2 summarizes the answers our survey provides to this question. What it shows is that Maryland nonprofit boards are fulfilling most of their major responsibilities but are nevertheless falling short in a number of areas. In particular:

- Two-thirds or more of Maryland's nonprofit boards are substantially or highly involved in the central management functions of their organizations--determining the organization's mission, developing its strategic plan, nominating its board members, deciding to undertake new activities, monitoring financial performance, deciding how to invest reserves, and setting the budget.

- Even here, however, nonprofit managers identify a need for improvement. Thus between 16 and 25 percent of the organizations indicated that they thought their boards should be *more* involved in determining the organization's mission, reviewing its financial performance, nominating board members, and establishing the budget; and nearly 30 percent indicated that they thought their boards should be *more* involved in developing the organization's strategic plan.

- In certain other key areas, board involvement is considerably less prominent. Thus only about half of the organizations indicated that their boards were substantially or highly involved in reviewing the organization's performance in achieving its mission, or setting its fundraising strategy; and *less than half* indicated that the boards had this level of involvement in hiring, firing, and evaluating the executive director, setting executive compensation, deciding to purchase goods or services from a board member, or setting policy on fees. Except for the setting of fundraising strategy, these were also areas where the respondents to our survey did not think the boards should be more involved.

- Board involvement in the evaluation of the organization's executive director seems to be an especially problematic area. In only half of the organizations had the board been involved in evaluating the work of the executive director in the previous two years, though this reached 81 percent in the case of the large agencies and 60 percent in the case of small and medium-sized agencies with some paid staff.

- Finally, even fewer organizations reported that their boards were substantially or highly involved in setting policies on fees (43 per-

> *Only about half of the organizations indicated that their boards were substantially or highly involved in reviewing the organization's performance, and less than half indicated that the boards were substantially involved in hiring, firing and evaluating the executive director.*

cent), seeking funds for the organization (38 percent), or deciding to bid on a government contract (25 percent). While the latter might be a matter appropriately left to professional staff, the former seem to be areas where board involvement is appropriate. Indeed, half of the agencies indicated that they thought their boards should be *more* involved in at least the fundraising tasks.

What these data indicate is that nonprofit boards in Maryland are taking a somewhat reserved posture with respect to their management functions. They tend to get involved in setting broad agency missions and strategies and in shaping the budget, but generally do not systematically monitor performance, evaluate key agency personnel, or get involved in many of the major types of decisions that have to do with the way the strategic goals are interpreted and carried out.

III. ACHIEVING THE MISSION: THE NONPROFIT SECTOR'S BOTTOM LINE

Quite apart from the role that boards of directors play, nonprofit organizations also potentially have more direct systems for ensuring effective management. Central to these is the measurement of success against a clearly articulated mission. The for-profit business world and the nonprofit sector differ substantially in this respect. For the business world, success is measured fundamentally in terms of profit. In the nonprofit world, by contrast, profit maximization is not available as a measure of success. Hence far more weight has to be placed on mission as the definition of the bottom line. Thus, the most fundamental question about the accountability of a nonprofit organization must be: *Is this organization achieving its mission?*

But how can organizations implement this nonprofit concept of a bottom line? There are two organizational practices that are fundamental:

• First, nonprofit organizations ought to have a clearly defined and articulated mission.

• Second, they should engage in regular efforts to track progress toward meeting their mission.

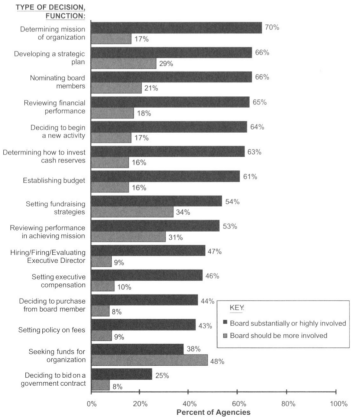

Figure 6.2 Actual vs. Desired Board Roles in Managing Maryland Nonprofit Organizations

Source: MANO/JHU-IPS Maryland Nonprofit Survey, 1996

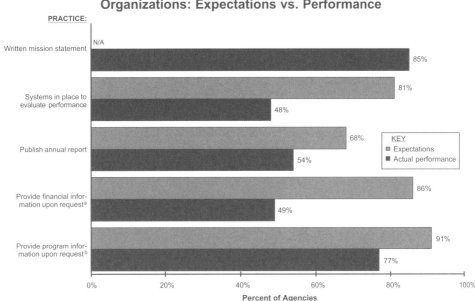

Figure 6.3 Management and Disclosure Practices of Maryland Nonprofit Organizations: Expectations vs. Performance

a Performance measure is percent of agencies that "regularly publish financial statements."
b Performance measure is percent of agencies that "regularly circulate program information."
Source: MANO/JHU-IPS Maryland Nonprofit Survey, 1996

Our survey indicates that the first of these two organizational practices is widely accepted, both in concept and in practice. Thus, as shown in Figure 6.3:

- The vast majority of nonprofits (85.1 percent) have written mission statements. Among medium-sized and large organizations, moreover, such statements are nearly universal (91.4 percent and 98.1 percent, respectively).

- This is consistent with the finding above that determining the mission of the organization is a major focus of the work of nonprofit boards.

The situation is not as favorable, however, when it comes to the evaluation of progress in implementing the mission:

- To be sure, the vast majority of respondents (81 percent) agreed that nonprofits "should have systems in place to evaluate the success of their programs and services."

- However, fewer than half of the organizations reported having systems in place to evaluate the success of *any* of their programs, and less than 20 percent reported having systems in place to evaluate *all* of their programs. Although, as Table 6.1 shows, large agencies are far more likely than small ones to have such evaluation systems in place, the generally limited assessment of organizational performance is still striking.

These data suggest another opportunity to strengthen accountability in Maryland's nonprofit sector. The majority of organizations need

> *Fewer than half of the organizations reported having systems in place to evaluate the success of any of their programs, and less than 20 percent reported having systems in place to evaluate all of their programs*

assistance to learn how best to incorporate performance measurement and evaluation into their day-to-day operations, and to involve their boards in the assessment of organizational activity.

IV. LIVING IN A GLASS HOUSE: THE DUTY TO DISCLOSE INFORMATION TO THE PUBLIC

Quite apart from whether they conduct their own evaluations, nonprofit organizations are also expected to provide information that will enable *others* to evaluate their operations. In fact, organizations in the nonprofit sector are subject to federal and state laws which require disclosure of a broad range of information. Basic tax documents, such as Form 1023 (the initial application for federal tax-exempt status) and Form 990 (the annual federal tax return), are required to be made available to members of the public upon request. Under state law, organizations which solicit charitable contributions must tell prospective donors that they have a right to obtain copies of financial statements and other documents.

These and other requirements are based on the notion that nonprofits are the beneficiaries of taxpayer-funded subsidies in the form of federal and state tax exemptions, tax deductions for charitable contributions, and, for some nonprofits, government contracts and grants. In this context, organizations in the nonprofit sector are expected to be accountable to the general public, or at least to the taxpaying public, for their actions. These disclosure requirements are, therefore, an accountability tool.

Our survey results demonstrate that nonprofit leaders generally recognize these disclosure obligations. Thus, as Figure 6.3 shows:

- Nine out of ten respondents agreed that nonprofit organizations should provide program information and nearly as many agreed that they should provide financial reports on request.

- Close to 70 percent also indicated that nonprofits should publish an annual report.

- In each case, moreover, the larger the agency, the more likely it is to endorse such reporting obligations.

Table 6.1
Share of Maryland Nonprofit Organizations
Reporting Various Disclosure Practices,
by Size of Agency

	Share of Agencies Reporting Indicated Practice				
	Conflict of interest policy in place	Publish an annual report	Regularly publish financial information	Regularly publish program information	Evaluation system in place for some or all services
Small (under $25,000)	11%	49%	42%	72%	39%
Medium ($25,000 to $1 million)	33	60	61	83	61
Large ($1 million and above)	62	74	73	93	86
ALL	21%	54%	49%	77%	48%

Source: MANO/JHU-IPS Maryland Nonprofit Survey, 1996

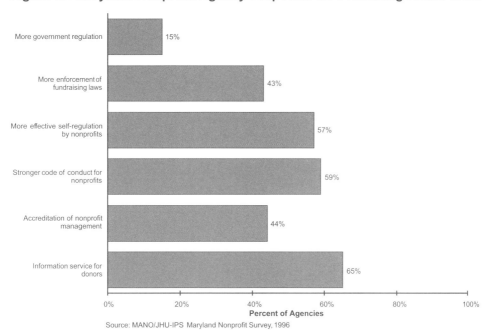

Figure 6.4 Maryland Nonprofit Agency Proposals for Promoting Public Trust

Source: MANO/JHU-IPS Maryland Nonprofit Survey, 1996

When it comes to actual performance, however, Maryland nonprofits fall considerably short of these standards. Thus, as Figure 6.3 shows:

- Only three-quarters of the agencies regularly circulate program information.
- Only 54 percent publish an annual report; and
- Less than half (49 percent) regularly publish financial statements.

For each of these, performance for the large agencies surpasses that for the small and medium agencies, as Table 6.1 shows. Nevertheless, these results suggest again that there is considerable room for improvement in the extent to which Maryland nonprofit organizations live up to their own expectations for ethical and accountable behavior. Despite their dependence on public confidence and support, many agencies still fall considerably short of the levels of communication and disclosure that they concede are desirable.

V. Promoting Public Trust in Maryland's Nonprofit Sector

Maryland nonprofit organizations, as a whole, thus adhere fairly well to the basic criteria of effective management evident in this field. At the same time, there remains room for improvement in the basic operations of these organizations so far as norms of ethics and accountability are concerned.

How do Maryland nonprofits respond to this "glass half-full or half-empty" situation? Are they inclined to leave things as they are or address the problems that exist?

The evidence from our survey suggests a willingness to tackle the gaps that exist and to work to improve public trust in the state's nonprofit sector. In particular, as noted in Figure 6.4:

- While few nonprofit leaders (15 percent) believe that there is a need for additional government <u>regulation</u>, a plurality (43 percent) believe that "there should be more vigorous enforcement of existing laws and regulations governing charitable solicitations."

- Strong support also exists for self-policing of the nonprofit sector. Thus, 57 percent of the respondents favor more effective self-regulation by nonprofits and 59 percent agree that the nonprofit sector should develop a strong code of conduct for nonprofit management practices.

- Similarly, 44 percent of the respondents supported the idea of an accreditation system for nonprofit management against only 18 percent who opposed such a system, and 65 percent endorsed the idea of developing "an information service to provide donors and prospective donors, upon request, with comparable information about charities."

CONCLUSION

Maryland nonprofit organizations thus adhere reasonably closely to the principles of good management in the nonprofit field. At the same time, by no means all of the organizations operate in accord with these principles. This is particularly true, moreover, among the small and medium-sized agencies. One reason for this may be the fiscal and other pressures under which this set of organizations is required to operate. Like other facets of nonprofit operations, the reporting and accountability procedures required to implement the management principles outlined here require resources--staff time, training, executive effort, and actual out-of-pocket costs.

From the evidence at hand, it seems clear that Maryland nonprofit agency managers generally understand and endorse the basic rudiments of ethical management outlined here. Given the importance that these practices hold for the preservation of public trust, and the vital importance that such trust holds for everything the sector tries to do, the time may be ripe to make sure that these principles are more fully put into effect.

CHAPTER SEVEN
NONPROFITS IN MARYLAND'S REGIONS

Like many states, Maryland is not so much a single entity as a collection of disparate regions, each with its own peculiar history, traditions, and economic structure. In population terms, the largest of these is the Baltimore region, which contains the City of Baltimore--a peculiar mix of Southern town and traditional northeastern industrial city-- and its surrounding, increasingly suburbanized, counties.[1] The second largest region is the set of counties that borders Washington, D.C., which includes high-income Montgomery County, middle-class Prince George's County, and the still somewhat rural Charles, Calvert, and St. Mary's counties. Different still are the two remaining regions--Western Maryland, a largely rural area with an Appalachian flavor; and the Eastern Shore, largely rural with a tradition of antebellum plantations.[2]

Previous chapters have commented in passing on the different characteristics of the nonprofit sector in these different regions. The purpose of this chapter is to pull these observations together and offer a more explicit profile of each region's nonprofit sector.

Because of data limitations, it is not possible to offer as fine-grained an analysis as regional variations might make desirable. We can, however, differentiate three broad divisions of the state: the Baltimore area, the Washington suburbs, and the balance of the state, with some differentiation between Baltimore City and the rest of the Baltimore area. Despite its limitations, however, this tripartite division does reveal some interesting differences within the state's nonprofit sector.

I. THE BALTIMORE REGION: TRADITIONAL CAPITAL OF MARYLAND'S NONPROFIT SECTOR

Overview

The Baltimore region, historically the hub of Maryland's economy and the center of its political power, is also the capital of its nonprofit sector. As Table 7.1 indicates, the Baltimore region houses half of the state's nonprofit institutions and accounts for close to two-thirds of its nonprofit employment. Of this total, Baltimore City alone accounts for over 38 percent of the employment, with the balance of the Baltimore area accounting for another 26 percent.

Economic Importance. Reflecting this, the nonprofit sector plays a significant part in the overall economy of this region. Close to 11 percent of all employment in the Baltimore area--one out of every 9 jobs--is in the nonprofit sector. This is proportionally one-fourth more than is true for the state as a whole. *In the City of Baltimore alone, nonprofit organizations account for 18 percent of all jobs, or nearly one in every 5.*

---

[1] For our purposes here, the Baltimore region includes the City of Baltimore and the following counties: Baltimore, Anne Arundel, Howard, Harford and Carroll.

[2] For purposes of this analysis, Western Maryland includes the following counties: Frederick, Allegany, Garrett, and Washington. The Eastern Shore includes these counties: Caroline, Cecil, Dorchester, Kent, Queen Anne's, Somerset, Talbot, Wicomico, and Worchester.

Table 7.1
Baltimore Area Nonprofits: A Profile

Characteristic	Baltimore Area[a]	Maryland	Baltimore as % of Maryland
Nonprofit Organizations, by Field	**6,289**	**12,570**	**50.0%**
Social services	20.5%	18.5%	110.8%
Education	19.8	16.9	117.2%
Culture, recreation	15.9	20.3	78.3%
Community development, environment	15.8	15.3	103.3%
Multipurpose	11.5	15.0	76.7%
Advocacy	5.4	3.1	180.0%
Other	4.3	4.7	96.5%
Health	2.8	2.2	127.3%
Employment, training	2.3	1.7	135.3%
Mental health	1.7	2.2	77.2%
Total	100.0%	100.0%	-
Nonprofit Employment, by Field	**118,410**	**185,088**	**64.0%**
Health	53.9%	52.8%	102.1%
Education	19.3	16.4	117.7%
Social services	16.6	19.5	85.1%
Culture, recreation	2.0	1.6	125.0%
Civic, social	2.5	3.4	73.5%
Other	5.7	6.3	90.5%
Total	100.0%	100.0%	
Nonprofit Share of Total Employment	**10.7%**	**8.5%**	**125.9%**
Nonprofit Employment Growth, 1989-96	**+19.3%**	**+24.2%**	**79.8%**
Size of Organizations			
Small	69.4%	66.6%	104.2%
Medium	23.4	26.7	87.6%
Large	7.2	6.7	107.5%
Total	100.0%	100.0%	-
Year of Formation			
1960 and earlier	26.7%	26.1%	102.2%
1961-1970	12.1	11.3	107.1
1971-1990	44.3	48.6	91.1
1991-present	16.8	14.1	119.1
Total	100.0%	100.0%	
Geographical Focus			
Neighborhood	34.2%	27.8%	123.0%
City or County	19.5	25.8	75.6
State or region	26.7	21.5	124.2
Multistate or wider	19.6	24.9	78.7
Total	100.0%	100.0%	-
% Poor-serving	**23.0%**	**16.2%**	-
Revenue Sources			
Private Giving	3.6%	4.4%	81.8%
Government	46.3	43.2	107.3
Earned Income	50.1	52.4	95.6
Total	100.0%	100.0%	-

Source: Employment data from Maryland Department of Labor, Licensing, and Regulation; Other data from MANO/JHU-IPS Maryland Nonprofit Survey, 1996.
[a] Includes Baltimore City, and Baltimore, Harford, Howard, Anne Arundel, and Carroll counties. Calculations may be subject to rounding error.

Declining Share. While the Baltimore area is the capital of the state's nonprofit sector, however, its relative position has been slipping in recent years. Between 1989 and 1996, for example, nonprofit employment in the Baltimore area increased by 19.3 percent, or 2.52 percent per year. During the same period, nonprofit employment in the state at large increased by 24.2 percent, an average increase of 3.02 percent per year. Nonprofit employment growth in the Baltimore area was thus only 80 percent of the state average.

Much of this relative lag in nonprofit growth in the Baltimore area is concentrated in the City of Baltimore. Between 1989 and 1996, nonprofit employment grew by only 8.9 percent in Baltimore City, as compared to 39.5 percent in the area outside the City. Clearly, as population has moved out of the central city into the suburbs, the nonprofit sector has followed suit. As a consequence, Baltimore City went from 44 percent of total nonprofit employment in the state in 1989 to 38 percent in 1996. Even so, the growth of nonprofit employment in Baltimore City stood in stark contrast to the overall decline in employment in the City. Between 1989 and 1996, the City of Baltimore lost 78,000 jobs. In this context, the 5,821 jobs added by the nonprofit sector made this sector one of the few real engines of local growth.

Composition

Reflecting its role as a regional capital of the nonprofit sector, the Baltimore area has disproportionate shares of:

- large agencies (7.2 percent vs. 6.7 percent for the state as a whole);
- state-wide or regional organizations (26.7 percent vs. 21.5 percent for the state as a whole);
- social service, education, health, and advocacy organizations; and
- nonprofit health, education, and cultural and recreational employment.

Close to 11 percent of all employment in the Baltimore area--one out of every 9 jobs--is in the nonprofit sector.

At the same time, reflecting the extensive neighborhood focus of many of the nonprofit organizations in Baltimore City, the area also boasts a disproportionate share of neighborhood-focused associations (34.2 percent vs. 27.8 percent for the state as a whole).

Client Focus

Given the presence of a sizable concentration of traditional urban problems in its central core, the Baltimore region's nonprofit sector has a considerably greater focus on the poor than is true of the state's nonprofit organizations generally. Thus 23 percent of the Baltimore area nonprofit organizations focus primarily on the poor, compared to only 16 percent for the state as a whole. In Baltimore City, moreover, this figure reaches 43 percent.

Revenue Structure

That Baltimore area nonprofit organizations have been able to concentrate more heavily on the poor, finally, is due in substantial part to their access to government support. Compared to the statewide averages, Baltimore area nonprofits receive relatively more of their support from government, and less from either private giving or earned income.

Summary

In short, the Baltimore area contains a rich mixture of traditional nonprofit organizations serving the cultural, educational, health, and social welfare needs both of Maryland's major urban center and of the state and region of which it is a part. Included here are the state's premier higher education institutions (Johns Hopkins, Goucher College, Loyola College), its leading hospitals and family service agencies, as well as its major cultural institutions (the Baltimore Symphony Orchestra, the Walters Art Gallery, the National Aquarium, and the Baltimore Opera). At the same time, Baltimore is home to a wide assortment of neighborhood, community development, employment and training, and anti-poverty agencies that reflect its ethnic heritage and its residue of serious urban ills. The region's nonprofit sector thus faces in two directions at once--toward the inner-city poor and toward the educated upper and middle classes. What is more, the region's nonprofit sector seems to be moving in two different directions--one toward the suburbanizing, middle class populations in the outlying counties; and the other toward the increasingly disadvantaged, heavily black, population of the urban core. In many cases, moreover, the same agencies are reaching out in both directions at the same time. In a curious way, therefore, the nonprofit sector may prove to be one of the best hopes for bridging the divide that increasingly characterizes these two separate communities.

II. THE WASHINGTON SUBURBS: BUILDING A CIVIC INFRASTRUCTURE

Overview

A very different nonprofit reality operates in Maryland's Washington suburbs. Originally developed as bedroom communities to the burgeoning Washington, D.C. metropolitan center, the commuter communities of Montgomery, Prince George's, and adjoining counties have increasingly developed their own civic infrastructures, including significant arrays of private, nonprofit institutions.

Though representing less than 6 percent of all jobs, nonprofit organizations accounted for 37 percent of all job growth in Maryland's Washington suburbs between 1989 and 1996.

As reflected in Table 7.2, about 30 percent of Maryland's nonprofit institutions are now located in the state's Washington suburbs. These institutions account for about one-fourth of total nonprofit employment in the state. This compares with about 36 percent of the state's population that resides in these counties.

Economic Importance. The nonprofit sector is thus still a "developing presence" in the Washington suburbs. Reflecting this, nonprofit employment is proportionally only two-thirds as important a part of the economy of these counties as it is to the overall state economy, constituting 5.8 percent of total employment as compared with 8.5 percent for the state as a whole, 11 percent (or proportionally nearly twice as great) in the Baltimore area, and 6.6 % in the Baltimore area outside of the City. Apparently, residents in these counties are able to rely on regional institutions located in Washington, D.C. or Baltimore for functions that indigenous nonprofit institutions might otherwise perform.

Table 7.2
The Nonprofit Sector in the Washington Suburbs of Maryland: A Profile

Characteristic	Washington Area[a]	Maryland	Washington as % of Maryland
Nonprofit Organizations, by Field	**3,631**	**12,570**	**28.9%**
Social services	14.7%	18.5%	79.5
Education	15.4	16.9	91.1
Culture, recreation	21.6	20.3	106.4
Community development, environment	16.2	15.3	105.9
Multipurpose	19.8	15.0	132.0
Advocacy	1.6	3.1	51.6
Other	8.0	4.7	170.2
Health	0.8	2.2	36.4
Employment, training	1.2	1.7	70.6
Mental health	<u>0.6</u>	<u>2.2</u>	<u>27.2</u>
Total	100.0%	100.0%	-
Nonprofit Employment, by Field	**44,003**	**185,088**	**23.8%**
Health	47.0%	52.8%	89.0%
Education	11.7	16.4	71.3%
Social services	25.8	19.5	132.3%
Culture, recreation	0.9	1.6	56.2%
Civic, social	5.4	3.4	158.9%
Other	<u>9.3</u>	<u>6.3</u>	147.6%
Total	100.0%	100.0%	
Nonprofit Share of Total Employment	**5.8%**	**8.5%**	**68.2%**
Nonprofit Employment Growth, 1989-96	**+37.6%**	**+24.2%**	**155.4%**
Size of Organizations			
Small	64.1%	66.6%	96.2%
Medium	31.9	26.7	119.5
Large	<u>4.0</u>	<u>6.7</u>	<u>59.7</u>
Total	100.0%	100.0%	-
Year of Formation			
1960 and earlier	24.3%	26.1%	93.1%
1961-1970	7.3	11.3	64.6
1971-1990	49.2	48.6	101.2
1991-present	<u>19.3</u>	<u>14.1</u>	136.9
Total	100.0%	100.0%	
Geographical Focus			
Neighborhood	20.8%	27.8%	74.8%
City or County	28.6	25.8	110.9
State or region	23.4	21.5	108.8
Multistate or wider	<u>27.2</u>	<u>24.9</u>	109.2
Total	100.0%	100.0%	-
% Poor-serving	**8.8%**	**16.2%**	**54.3%**
Revenue Sources			
Private Giving	7.1%	4.4%	161.4%
Government	12.4	43.2	28.7
Earned Income	<u>80.5</u>	<u>52.4</u>	<u>153.6</u>
Total	100.0%	100.0%	

Source: Employment data from Maryland Department of Labor, Licensing, and Regulation; Other data from MANO/JHU-IPS Maryland Nonprofit Survey, 1996.
[a] Includes Montgomery, Prince George's, Charles, Calvert, and St. Mary's Counties.
Calculations may be subject to rounding error.

A Growing Presence. If the nonprofit sector is still a relatively smaller component of the economy in the Washington suburbs than elsewhere in the state, however, it is clearly a growing presence. Nonprofit employment in Maryland's Washington suburbs grew by 37.6 percent between 1989 and 1996 compared to 24.2 percent for the state as a whole. This translates into an average annual increase of 4.6 percent per year for the Washington suburbs, compared to 3.0 percent for the state as a whole. What is more, nonprofit job growth in this area far exceeded the growth of jobs generally. Indeed, though representing 4.4 percent of all jobs as of 1989, nonprofit organizations accounted for 37 percent of all *job growth* in Maryland's Washington suburbs between 1989 and 1996. Clearly, this sector is a "comer" in the economy of this region.

Composition

Reflecting its less highly developed character, the nonprofit sector in Maryland's Washington suburbs also differs in composition from that in the Baltimore region and the state at large. In particular, this region has:

- A larger share of young agencies than the state as a whole (19.3 percent vs. 14.1 percent);

- A correspondingly smaller share of very large agencies (4.0 percent vs. 6.7 percent for the state as a whole);

- Significantly smaller shares of nonprofit employment in the traditional fields of health, education, and culture and recreation, where nearby Washington, D.C. probably offers service options;

- Significantly higher shares of nonprofit employment in the fields of social services and civic and social activity. The former of these very likely reflects the heavy reliance on day care and family services in this heavily two-worker-family area.

- Less of a neighborhood focus and more focus on statewide, regional, and national or international concerns.

Funding Base

Reflecting its "middle class" orientation, the nonprofit sector in the Washington suburbs also seems to have a distinctive funding base compared to its counterparts in Baltimore and elsewhere in the state. In particular, as Table 7.2 also shows, over 80 percent of the income of suburban Washington nonprofit organizations comes from service fees and charges compared to 52 percent for the state as a whole. By contrast, a much smaller share comes from government (12.4 percent vs. 43.2 percent for the state). Evidently the nonprofit organizations in the Washington suburbs have less access to governmental support than is the case for their counterparts elsewhere in the state. To the extent they have been able to offset this lagging public support, moreover, they have done so mostly through turning to fees and charges rather than private charitable support. They are consequently even more oriented to a paying clientele than is the case for nonprofits elsewhere in the state.

Client Focus

Very likely reflecting this funding situation, the nonprofit sector in Maryland's Washington suburbs seems even less focused on those in poverty than is the case for the rest of the state. Thus, only about 9 percent of the agencies in this region reported focusing primarily on the poor, compared to 16 percent for the state as a whole. While this may also be a byproduct of the generally middle class character of the populations in these counties, the fact is that significant pockets of poverty are also present in this area. Yet, whether by choice or necessity, relatively few local nonprofit organizations have been able to focus very exclusively on these poverty problems.

Summary

The Washington suburbs of Maryland are thus in a developmental posture so far as the nonprofit sector is concerned. Nonprofit employment has been growing rapidly in this area--far more rapidly than the economy as a whole--yet its overall scale still lags behind that in the rest of the state in relative terms. Though developing indigenous nonprofit institutions to promote social and civic purposes and provide the social services associated with a population that contains many two-earner households with children, this area still apparently relies on the urban core of Washington for much of the health, education, and cultural services that nonprofits traditionally provide. The result is a more "commercial," middle-class-oriented nonprofit sector that gets most of its income from fees and charges, has less involvement with government, and tends to be less focused on the poor than is the case with nonprofits in the state as a whole.

III. WESTERN MARYLAND AND THE EASTERN SHORE

Overview

Though often associated exclusively with urban areas, the nonprofit sector is also very much in evidence in the more rural parts of Maryland in the western counties and the Eastern Shore. With 15 percent of the state's population, these areas account for just over 12 percent of its nonprofit employment and around 20 percent of its nonprofit organizations, as reflected in Table 7.3.

Economic Importance. Generally speaking, the nonprofit sector tends to be more volunteer-oriented in these two regions than in the state as a whole. While accounting for 12 percent of all nonprofit employment, therefore, these regions absorb more than 20 percent of the volunteer effort. Nevertheless, nonprofit organizations are also significant employers in these two regions. In fact, 7.6 percent of all jobs in Western Maryland and the Eastern Shore are in the nonprofit sector. This represents half as many jobs as are employed in all the manufacturing industries in these regions.

Recent Growth. The nonprofit presence in these areas has been growing significantly in recent years, moreover. Thus, between 1989 and 1996, nonprofit employment grew by close to 28 percent. By comparison, during this same period total employment in these regions grew by only 14 percent. The nonprofit sector was thus a disproportionate contributor to job growth.

Composition

The composition of the nonprofit sector in Western Maryland and the Eastern Shore resembles that in the state as a whole, with the following exceptions:

- Health providers seem more heavily represented among the nonprofit institutions in these areas than they are statewide. This may reflect the isolated character of these areas and the consequent need for local health institutions.

- Nonprofit educational institutions are much less prominent. More of the educational burden in these areas is therefore borne by public institutions, both at the elementary and secondary level and at the higher education level;

> *Though often associated exclusively with urban areas, the nonprofit sector is also very much in evidence in the more rural parts of Maryland in the western counties and the Eastern Shore.*

Table 7.3
Nonprofit Organizations in Western Maryland and the Eastern Shore: A Profile

Characteristic	Western MD & Eastern Shore[a]	Maryland	W.MD and E. Shore as % of State
Nonprofit Organizations, by Field	**2,650**	**12,570**	**21.1%**
Social services	17.9%	18.5%	96.8
Education	12.2	16.9	72.2
Culture, recreation	29.6	20.3	145.8
Community development, environment	13.1	15.3	85.6
Multipurpose	17.2	15.0	114.7
Advocacy	-	3.1	0
Other	1.4	4.7	29.8
Health	2.4	2.2	109.1
Employment, training	0.3	1.7	17.6
Mental health	5.9	2.2	268.2
Total	100.0%	100.0%	-
Nonprofit Employment, by Field	**22,675**	**185,088**	**12.3%**
Health	58.5%	52.8%	110.8%
Education	10.6	16.4	64.6%
Social services	22.1	19.5	113.3%
Culture, recreation	0.6	1.6	37.5%
Civic, social	4.3	3.4	126.4%
Other	3.9	6.3	61.9%
Total	100.0%	100.0%	-
Nonprofit Share of Total Employment	**7.6%**	**8.5%**	**89.4%**
Nonprofit Employment Growth, 1989-96	**+27.6%**	**+24.2%**	**114.0%**
Size of Organizations			
Small	64.7%	66.6%	97.1%
Medium	26.9	26.7	100.7
Large	8.4	6.7	125.4
Total	100.0%	100.0%	-
Year of Formation			
1960 and earlier	26.5%	26.1%	101.5%
1961-1970	14.1	11.3	124.8
1971-1990	58.2	48.6	119.8
1991-present	1.2	14.1	8.5
Total	100.0%	100.0%	
Geographical Focus			
Neighborhood	23.0%	27.8%	82.7%
City or County	37.9	25.8	146.9
State or region	4.5	21.5	20.9
Multistate or wider	34.6	24.9	12.8
Total	100.0%	100.0%	-
% Poor-serving	**10.0%**	**16.2%**	**61.7%**
Revenue Sources			
Private Giving	7.0%	4.4%	159.1%
Government	50.3	43.2	116.4
Earned Income	42.7	52.4	81.5
Total	100.0%	100.0%	

Source: Employment data from Maryland Department of Labor, Licensing, and Regulation; Other data from MANO/JHU-IPS Maryland Nonprofit Survey, 1996.
[a]Includes Frederick, Allegany, Garrett, Washington, Caroline, Cecil, Dorchester, Kent, Queen Anne's, & Somerset counties. Calculations may be subject to rounding error.

- A disproportionate share of the nonprofit employment in these areas is in the social services field (22.1 percent vs. 19.5 percent);
- While there are proportionally more nonprofit cultural and recreational associations in Western Maryland and the Eastern Shore, employment in these types of organizations is proportionally smaller. This is very likely because recreational associations--e.g. sports clubs--predominate in these regions and these tend more commonly to be volunteer based, as opposed to the more professional cultural institutions common in urban areas.
- Organizations are more likely to have a countywide focus instead of a neighborhood focus, though the presence of a number of multi-state organizations constitutes an anomaly perhaps explainable by the tendency of organizations in these far reaches of the state to serve people in neighboring states as well.

Funding Base

Perhaps the most distinctive feature of the nonprofit sector in Western Maryland and the Eastern Shore is its funding base. Compared to the state as a whole, government plays a significantly greater role in these regions and earned income a significantly smaller role. In addition, private giving is proportionally twice as important in these two regions as in the state generally, though it still accounts for only 7 percent of total income.

Client Focus

The nonprofit sector is also not focused exclusively on the poor in these areas. As Table 7.3 shows, only 10 percent of the agencies report that the poor comprise half or more of their clientele. Here, as well, therefore, nonprofit organizations tend to serve a broad cross-section of the community.

Summary

In short, a sizable nonprofit sector operates even in the more outlying regions of Maryland. These organizations are actively engaged in the provision of basic services in such fields as health and social services, and are available to serve a variety of recreational and civic purposes as well.

CONCLUSION

If Maryland is a state of disparate regions, it is also a state of disparate nonprofit sectors. Nonprofit institutions have emerged to meet the varied needs of different population groups in each of the state's regions. As a consequence, this sector takes a variety of different forms, with different constellations of services and different patterns of service delivery and finance.

Despite these differences, however, perhaps the most striking feature of the state's nonprofit sector is its ubiquity. Whether in the urban core of Baltimore or the hilly counties in the western portion of the state, Marylanders have come to recognize the vital importance of nonprofit organizations to meet their pressing needs and enrich the quality of their social life. While the state's nonprofit sector has many different features, it also shares many common characteristics, and many common concerns. While celebrating the variety of local nonprofit life, it is therefore important not to lose sight of the things the state's nonprofit organizations also share in common, and the priorities around which they may consequently usefully coalesce. It is to these matters that we therefore now turn.

CHAPTER EIGHT
CONCLUSIONS AND RECOMMENDATIONS

Maryland's nonprofit sector emerges from the analysis in this report as the "Rodney Dangerfield" of Maryland society. This set of organizations provides crucial services to virtually every man, woman, and child in the state--health care, family counseling, day care, nursing care, home health services, employment and training, education, emergency aid, music, dance, theater, art, and recreation. In the process, it contributes massively to the quality of community life. It does so, moreover, not only in the traditional urban core but also in the suburban counties surrounding Baltimore and Washington D.C. and in the rural reaches of Western Maryland and the Eastern Shore.

Not only that, but this set of organizations turns out to be one of the principal engines of the state's economy. More people are employed by nonprofit organizations than by all of the state's manufacturing businesses, all of its construction businesses, all of its transportation and communications businesses, and all of its finance, insurance, and real estate businesses. What is more, this sector has been one of the major sources of employment growth in the state in recent years. Indeed, half of the net gain in employment that the state achieved between 1989 and 1996 was provided by the state's nonprofit organizations.

Despite their importance, however, nonprofit organizations attract little attention, let alone respect. Annual newspaper summaries of the state's economy barely acknowledge the sector's existence. The state's economic statisticians overlook this sector altogether and its Department of Business and Economic Development rarely focuses on its needs.

Maryland's nonprofit sector emerges from the analysis in this report as the "Rodney Dangerfield" of Maryland society.

No wonder popular understanding of this sector is so limited and inaccurate. According to widespread beliefs, for example, the nonprofit sector is principally financed by voluntary contributions. In fact, as the data reported here demonstrate, its major source of revenue is earned income followed closely by government support. Similarly, popular conceptions picture this set of organizations as principally serving the poor when in fact most agencies serve the broad middle class. Indeed, far from moving closer to these popular conceptions, the data reported here show that the Maryland nonprofit sector seems to be moving in the opposite direction--away from the inner cities and toward the suburbs, away from the poor and toward the middle class, away from reliance on private charity and toward reliance on fees and charges. One illustration of this is the significant decline in the share of nonprofit income coming from private charity that seems to have occurred since a comparable study was carried out just ten years ago. At that time, private giving constituted 20 percent of the income of nonprofit human service agencies other than hospitals and higher education institutions in the Baltimore region.[1] As of 1996, this figure stood at 10 percent for the state as a whole.

While these disparities between fact and fiction might be a matter of only academic interest in other times, they have become matters of pressing practical concern at the present because they seem to be leading to counterproductive policies and a serious decline in public confidence in this crucial set of community institutions. In a sense, the nonprofit sector is being hoisted on its own mythology: convinced that these organizations rely principally on private philanthropy, the public willingly endorses cutbacks in government support without recognizing the implications for the ability of nonprofit organizations to do their work. Then, not comprehending the fiscal strain nonprofits are experiencing as a consequence the public reacts negatively to reports of nonprofit involvement in commercial activity to survive. Finally, to complete the circle, nonprofits are forced to react to this situation by reducing their own investments in staff development and evaluation and by cutting back on mission-related activities such as free service to the poor, thereby threatening the quality of their own programs and eliminating the very activities that give them their distinctive character.

THE NEED FOR RENEWAL: A BLUEPRINT FOR ACTION

To cope with this situation, a thoroughgoing process of renewal is needed both within the nonprofit sector and between it and the public at large. This will involve some basic rethinking of the current direction of nonprofit evolution in the state and a serious program of action to reclaim the sector's future.

While the full development of such a program will require an extensive process of consultation and debate, some of the major areas requiring immediate attention are already visible from the findings in this report. Five of these in particular deserve mention here.

1. ENGAGING OPINION LEADERS: A MARYLAND CIVIL SOCIETY COMMISSION

First and foremost, a process is needed to engage opinion leaders in Maryland in a serious re-evaluation of the state's nonprofit sector and the directions in which it is headed. A set of institutions as vital to the social and economic health of this state should no longer be ignored so totally by the state's political, economic, and social leadership.

First and foremost, a process is needed to engage opinion leaders in Maryland in a serious re-evaluation of the state's nonprofit sector and the directions in which it is headed.

Accordingly, the Governor, in conjunction with leading business and philanthropic institutions, should appoint a *Maryland Civil Society Commission* to initiate a process of public inquiry into the role and contributions of the nonprofit sector and the ways in which this sector might be strengthened. The work of this Commission should be funded by a combination of government and philanthropic sources and resources should be made

---

[1] Lester M. Salamon, *More Than Just Charity* (Baltimore: Johns Hopkins Institute for Policy Studies, 1990), p. 38.

available not only to conduct the Commission's business, but also to explain the nonprofit sector to the state's population through effective paid advertising.

As part of its deliberations, this Commission could usefully consider the threats being posed to the nonprofit sector by shrinking government support, by efforts in some jurisdictions to eliminate the property tax exemptions nonprofit organizations now receive, and by proposals to limit the advocacy activity of nonprofit organizations. These threats often grow out of insufficient understanding of the role that nonprofit organizations play and the contribution they make.

> *No business sector would stand for the gross lack of basic information that the nonprofit sector has had to endure in this state.*

The Commission could usefully sponsor inquiries into key facets of the operation of the nonprofit sector in Maryland and hold hearings in different parts of the state to review the status of Maryland's "civil society organizations" and the steps that could be taken to strengthen this crucial facet of Maryland community life. At the end of its work, the Commission could produce a "white paper" on the future of Maryland's nonprofit sector that could then be used as a focus for community forums throughout the state. The result could be a new appreciation of the role of nonprofit organizations in the social and economic life of this state, a new awareness of the threats being posed to this set of institutions, and a new basis for cooperation among government, business, and nonprofit groups in the solution of public problems. This would not only have immediate social payoffs, but also, by promoting a strong "civic culture," help attract additional businesses to the state.

2. MONITORING THE HEALTH OF THE STATE'S NONPROFIT SECTOR

"Out of sight, out of mind" has historically been the posture of economic statisticians toward the nonprofit sector both nationally and in the separate states. The only reliable economic statistics available on the nonprofit sector are those produced once every five years by the U.S. Census Bureau through its *Census of Service Industries* and those generated by the Internal Revenue Service from the Form 990 information filings submitted each year by registered nonprofit organizations. Neither of these sources is adequate, however. The Census data cover only organizations with at least one paid employee, frequently vary in their coverage, and are usually reported at the statewide or metropolitan area level only. The IRS data suffer from serious error rates because IRS does not monitor the reporting forms closely and reporting is somewhat haphazard. Beyond this, however, there is no entity with the responsibility to analyze these data for the State of Maryland. The State's Department of Labor, Licensing, and Regulation, which regularly reports on employment and payrolls by industry does not break out nonprofit employers for separate reporting. The Secretary of State collects information on organizations that solicit contributions in the state, but this covers only a small fraction of organizations and the data files are imperfect, to say the least, in both coverage and content.[2]

Given the scale and importance that the nonprofit sector has attained, it seems clear that this situation is no longer tenable. No business sector would stand for the gross lack

---

[2] As noted above, according to the Secretary of State's files, there are 3,310 charities operating in Maryland, of which 2,389 have Maryland addresses. According to the Secretary of State's files, these organizations have revenues of $7.186 billion. By contrast, the Internal Revenue Service records close to 13,000 501(c)(3) and 501(c)(4) organizations in Maryland and our estimate puts their revenues in excess of $12 billion.

of basic information that the nonprofit sector has had to endure in this state (and others). A major effort is therefore needed to create a regular series of data on the nonprofit sector in Maryland, and a process to analyze these data on a regular basis.

Fortunately, in the course of this work, we discovered two data sources that could contribute massively toward such a goal. One of these is the monthly employment and wage data collected by the Maryland Department of Labor, Licensing, and Regulation as part of its regular economic data series. In response to our requests, the Department's Division of Labor Market Analysis and Information was able to break nonprofit employers out separately in its data and provide these data to us. This should now be done on a regular basis so that the Department's regular monthly and quarterly reports on employers, employees, and payrolls in the state separately identify the numbers of nonprofit employers, employees, and payrolls by county and field. This simple change could vastly improve our ability to track the nonprofit sector in the state, permitting us to do the kind of analysis that was offered in Chapter Two above on a regular basis.

Similarly, the state's Comptroller's Office is capable of generating highly useful information on charitable giving in the state based on state income tax returns. While some of this information is published in the office's annual *Statistics of Income* publication, more detailed information should be presented on an annual basis.

With these data made available, the state government and the state's philanthropic community should join forces to commission an annual "State of Maryland's Nonprofit Sector" report. Such a report could be featured along with existing reports on the business sector in annual round-ups on the state's economy.

3. BOOSTING PRIVATE GIVING AND BUTTRESSING THE STATE'S PRIVATE PHILANTHROPIC BASE

One of the more startling findings of this report is the relatively limited support that nonprofit organizations in this state receive from private charitable giving, and the much greater reliance that is placed on earned income from fees and charges. Indeed, as noted above, the extent of reliance on private charitable support seems to be shrinking rather than growing. This is all the more significant, moreover, in view of expectations on the part of the majority of nonprofit organizations that private giving will grow enough to offset stagnant or declining governmental support and meet the increasing demand for nonprofit services. Under present circumstances, these expectations seem totally unrealistic. If no other action is taken, therefore, declining government support will lead to even greater reliance on commercial income instead, which could further blur the lines between nonprofit and for-profit activity.

The state government and the state's philanthropic community should join forces to commission an annual "State of Maryland's Nonprofit Sector" report.

Bolstering the levels of private giving and buttressing the state's private philanthropic base are crucial to the continued vitality and independence of the state's nonprofit sector.

While it is not reasonable to think of private giving as a substitute for government funding, the fact is that bolstering the levels of private giving and buttressing the state's private philanthropic base are crucial to the continued vitality and independence of the state's nonprofit sector. How might this be done?

Fundamentally, a multi-pronged approach will be required. But four steps seem particularly promising:

(i) Improved Tax Incentives for Giving

In the first place, the incentives for charitable giving should be increased, particularly for the two thirds of all taxpayers who do not itemize their tax deductions. Under present circumstances, such taxpayers have little financial incentive to give to charity. While such incentives are hardly the only reason why individuals make charitable deductions, past research, including a recent study by the accounting firm of Price Waterhouse,[3] demonstrate convincingly that the availability of tax incentives affects the size of gifts in a very positive way.

The major tax-related incentives for charitable giving operate, of course, through the federal income tax. Nevertheless, there are also things that could be done at the state level. By taking such steps, Maryland could send a strong signal about its special commitment to private philanthropy and citizen action. More concretely, two measures are worthy of consideration:

- **An Above-the-Line Charitable Deduction for Nonitemizers.** As a first step, non-itemizers who contribute more than a given amount to charity could be permitted to claim deductions for these contributions even though they take the standard deduction. Such a provision was in place nationally for a limited period in the early 1980s and then phased out. Properly designed, such a provision could help give a much broader array of state citizens a stake in the health of the state's nonprofit sector.

- **A Charity Tax Credit.** Even more effective than an above-the-line charitable deduction for non-itemizers would be the adoption of a "tax credit" for charitable contributions along the lines of one recently adopted by North Carolina. A "tax credit" is a more powerful incentive than a tax deduction because it delivers benefits in the form of direct reductions of tax liabilities rather than indirectly in the form of reductions in taxable income. A tax credit worth 5 percent of all contributions over a set amount (e.g. 2 percent of taxable income) could serve as a powerful incentive for increased charitable contributions in the state.

(ii) A Community Foundation Initiative

One of the more effective vehicles for charitable fundraising being used increasingly in this country and elsewhere in the world is the community foundation. Community foundations are institutions that assemble charitable endowments from numerous donors and manage them in a flexible and responsive way to address community needs. In communities such as Cleveland, New York, Chicago, and San Francisco, community foundations have become major philanthropic institutions able to bring significant resources to bear on local problems and serve as a catalyst to encourage the development of the local nonprofit sector.

Fortunately, a number of community foundations have been established in this state--in Baltimore, Montgomery County, Prince Georges County, Columbia, Frederick County, and the Eastern Shore. Aside from the Baltimore Community Foundation, however, these institutions are fairly young and public recognition remains quite limited. What is more, the existing organizations have not joined forces in a meaningful way to promote the idea of community foundations at the statewide level or to assist the state's nonprofit sector in a coherent fashion.

---

[3]Price Waterhouse and Caplin and Drydale, Chartered, *Impact of Tax Restructuring on Tax-Exempt Organizations* (Washington, DC: The Council on Foundations and Independent Sector, 1997).

Given the emergence of sizable nonprofit sectors throughout Maryland, the time may be ripe for a major push to encourage the development of these community-based philanthropic institutions. Beyond this, it may be useful to consider creating a statewide community foundation, or a coordinating mechanism among the existing foundations, to elevate the profile of the community foundation phenomenon, to organize statewide programs, and to provide a stronger source of philanthropic leadership in the state. The "nonprofit renewal" project outlined here could provide a useful, neutral focus for such a statewide community foundation initiative.

(iii) Fundraising Training

Continued attention also needs to be given to strengthening the fund raising of Maryland's nonprofit organizations. In the years since *More Than Just Charity* was published, a number of new programs have been created by the Maryland Association of Nonprofit Organizations ("Maryland Nonprofits") and others. These efforts need to continue and be bolstered by additional resources from foundations and government funding sources. What is more, as noted below, they need to be more explicitly related to, and possibly integrated into, a broader program of training for nonprofit managers, volunteers, and board members since fundraising cannot be effective if conducted out of the context of organizational management and strategy more generally.

(iv) Celebrate Giving and Partnership

Finally, the relatively limited private giving in Maryland suggests the need for a major *public relations effort* to promote increased charitable giving. Such a campaign should be financed by both public and private sources. Its goal, moreover, should not simply be to trumpet the vital need for charitable giving, but also to acknowledge the importance of cooperation among government, business, and the nonprofit sector in addressing the state's problems and enhancing its quality of life. Such a campaign would help foster a sense of civic involvement, which can be one of the state's most potent economic development attractions.

4. BUILDING ORGANIZATIONAL CAPACITY

A fourth area that needs continued attention in Maryland is improvement in the basic organizational capacity of the state's nonprofit sector. As in other states, public and private funders in Maryland tend to take a project orientation to their work--they support particular programs and projects but do not give much attention to the capacity of the organizations that are supposed to deliver the programs or perform the projects. To the contrary, organizational capacity is treated as unwanted "overhead" that diverts resources from the important work of delivering services. The findings of a recent national study thus ring true in the context of Maryland: "Foundations' attitudes have long encouraged nonprofit organizations to focus on mission and to regard organizational capacity as worthwhile in principle but a distracting burden in practice."[4]

> *The relatively limited private giving in Maryland suggests the need for a major public relations effort to promote increased charitable giving and emphasize the importance of nonprofit partnerships with government and business to solve community problems.*

---

[4]Christine Letts, William Ryan, and Allan Grossman, "Virtuous Capital: What Foundations Can Learn from Venture Capitalists," *Harvard Business Review* (March/April 1997), p. 37.

A similar point could be made about government funders as well. *As a consequence, no one is effectively investing in nonprofit organizational capacity.* This differs rather markedly from the way venture capitalists operate in the corporate sector--there building an effective organization is viewed as crucial to the success of a new venture.

The creation of the Maryland Association of Nonprofit Organizations in response to a suggestion made in our earlier report, *More Than Just Charity,* has usefully corrected this problem in part in Maryland. Maryland Nonprofits has succeeded in elevating the nonprofit sector within the state's decision making councils and has contributed importantly to skills upgrading within the sector. *But it is now time to build on that foundation and extend it by following up on other suggestions in that report and going beyond them.* In particular, at least four steps are needed:

(i) Training

In the first place, Maryland still lacks an accessible, full-fledged program of nonprofit management training comparable to those that have developed in other states. A significant program of short-term training seminars has now been organized by Maryland Nonprofits, and this program should be continued and expanded.

At the same time, however, a more intensive degree program and a related "certificate program" are also needed to give the in-depth knowledge and theoretical grounding that an industry of this scale and complexity requires. A group of nonprofit leaders is now working with the Johns Hopkins Institute for Policy Studies to develop such a degree program. These efforts deserve widespread support.

Skills upgrading on a regular basis has become crucial in all parts of the economy, and the effectiveness of the nonprofit sector depends on it no less.

(ii) Technical Assistance: A Nonprofit Management Improvement Fund

Training by itself will not, however, solve all of the management problems being faced by the state's nonprofit organizations. In addition, more intensive, yet affordable, technical assistance services are needed in areas where one-to-one contact is the most appropriate vehicle for learning. Nonprofit organizations need to rethink basic features of their organizational structure and behavior just as business organizations have recently done. Such "re-engineering" is not without its costs, however. To cover them, a *Nonprofit Management Improvement Fund* could usefully be established among local funders. Such a fund could be administered by the new coalition of community foundations mentioned above, though Maryland Nonprofits would be another possible agent. Nonprofit organizations could then apply to this fund for support for management training for their staff or for more basic organizational analysis and overhaul. Funds would be available to cover staff training, consulting services and, where appropriate, the fixed costs of organizational restructuring.

(iii) Improved Benefits

One of the significant problems that nonprofit organizations reported in our survey was the difficulty in maintaining adequate benefit levels for agency staff. One reason for this may be the constraints imposed on nonprofit compensation and benefit levels by government

> *Maryland still lacks an accessible, full-fledged program of nonprofit management training comparable to those that have developed in other states.*

contracts, which often make inadequate provision for these items. In addition, many small nonprofits lack access to benefit plans that can give them the best rates on health, retirement, and unemployment coverage.

To remedy this set of problems, several steps seem appropriate:

- First, the Governor's Council on Management and Productivity should examine the reimbursement provisions under contracts with nonprofit providers to be sure that benefit allowances are adequate.

- Second, Maryland Nonprofits should be encouraged to continue its successful efforts to develop benefit packages that can help smaller nonprofits gain access to crucial benefits at reasonable cost.

- Finally, where appropriate, nonprofit employees should be offered access to state employee benefit plans for health and related benefits. Such a move would take account of the public-service role that nonprofit organizations perform and relieve small organizations of the costly burden of developing their own benefit package arrangements.

(iv) Strengthened Infrastructure Organizations

Finally, as part of a strategy of institutional strengthening, further efforts should be made to develop Maryland Nonprofits into an effective service and advocacy vehicle for the state's nonprofit sector. Larger nonprofits that have been hesitant to join this association should be encouraged to do so. Funders that have preferred to finance programs while neglecting the infrastructure of "their" sector should change this attitude and provide core support to this crucial institution. What is more, Maryland Nonprofits should be equipped to broaden the public information and advocacy activities in which it is engaged. Maryland Nonprofits has clearly demonstrated its value to the Maryland nonprofit community and to the state more generally. It is now time for the Maryland nonprofit and philanthropic community to demonstrate it values what this organization has accomplished, and what it promises to do in the future.

5. Promoting Public Confidence

The future viability of Maryland's nonprofit sector, and its ability to tap additional private support, ultimately depend on the level of public confidence and trust. Yet that trust has been shaken in recent years by disclosures of fraud and abuse among some national nonprofit organizations, and by a general questioning of the effectiveness of social programs of any sort.

In view of this, serious steps need to be taken to recapture the public's trust. For the most part, this will require action on the part of the nonprofit sector itself. While increased government regulation or enforcement activities may be helpful to ferret out outright fraud, such activities merely redress wrongdoing and do little to create a positive image of nonprofits. Ultimately, therefore, nonprofits must demonstrate that they are worthy of public support. This can be done in at least two ways.

- **Performance Measurement.**
  In the first place, nonprofit organizations must become more explicit and self-conscious in measuring the results of their activities.

> *As a part of a strategy of institutional strengthening, further efforts should be made to develop Maryland Nonprofits into an effective service and advocacy vehicle for the state's nonprofit sector.*

This will require, as a prerequisite, a clearer articulation of missions, better appreciation of how various activities contribute to those missions, and a sense of how effectiveness will be determined. This must then be married to systematic efforts to measure results. None of this will be possible, however, without support. Foundations and government agencies that finance nonprofit activity should therefore build evaluation components more squarely into their grants so that nonprofit organizations can generate the information that is needed to demonstrate the worth of what they do.

- **An Ethics and Accountability Campaign.** Secondly, nonprofits should, collectively, articulate standards for appropriate organizational behavior in the nonprofit sector and create a mechanism to encourage organizations to adhere to these standards. The Maryland Association of Nonprofit Organizations has made important progress along these lines by formulating a code of conduct for nonprofit organizations in Maryland. This code now needs to be endorsed by the state's nonprofit organizations and a meaningful system put in place to encourage adherence to it. This could best be done by a neutral entity that can certify that organizations are operating in a way that is consistent with what the nonprofit code of conduct suggests. Funding for such an entity could come in part from taxes on the unrelated business activities of nonprofit organizations. In this way, the limited tax revenues that nonprofit organizations generate could be used to strengthen the state's nonprofit sector and the level of public confidence in it.

CONCLUSION

The State of Maryland has a vast resource for good in its nonprofit sector. The organizations that comprise this sector provide vital community services and contribute to the quality of life in dozens of other ways.

As this report has shown, moreover, these organizations play an important part in the state's economy as well. Indeed, they have become one of the state's principal engines of job growth.

To date, however, the role of this important set of organizations has been systematically overlooked. Worse yet, that neglect has recently begun to take its toll. This is evident in significant popular disaffection, limited levels of private charitable support, significant threats to organizational effectiveness, and a slow, steady drift towards greater commercialization.

The objective of this report has been to bring this important sector out of the shadows, to document its basic scale and contours, and to identify some of the challenges it now confronts. The ultimate goal, however, is not simply to give this sector the attention it deserves, but to stimulate the actions that will allow it to achieve the promise of which it is capable. Hopefully, that task can now proceed with greater vigor.

> *Nonprofit organizations must become more explicit and self-conscious in measuring the results of their activities.*

> *The ultimate goal of this report is not simply to give Maryland's nonprofit sector the attention it deserves, but to stimulate the actions that will allow it to achieve the promise of which it is capable.*

APPENDIX A
MARYLAND NONPROFIT SURVEY METHODOLOGY[1]

This Appendix describes the methodology used for the survey of Maryland nonprofit agencies that forms the basis for much of the discussion in Chapters 3-7 of this report. The discussion focuses first on the sampling methodology that was used, then examines the representativeness of the sample that resulted, turns next to the handling of missing or inconsistent data, and concludes with a discussion of the weighting system that was used to blow up the sample responses to the known population of agencies.

1. SAMPLING METHODOLOGY

1.1 Sampling Frame

No completely reliable listing of private, nonprofit organizations exists for the State of Maryland (or any other state). Records kept by the Secretary of State cover only organizations that actively solicit contributions from the public in Maryland, and even then only those that raise $25,000 or more in a single year. Nonprofit organizations generally secure a tax exempt certificate from the Internal Revenue Service, which maintains an Exempt Organization Master File based on these exemption records and on a special reporting form (Form 990) that registered nonprofit organizations are required to file each year. However, there is no legal obligation on nonprofits to secure such a formal exemption and many organizations apparently do not do so. What is more, the Internal Revenue Service special reporting form (Form 990) is only required of organizations with $25,000 or more of income. Beyond this, organizations operating in Maryland may be branches of national or regional organizations that file with the IRS from their headquarters office and thus get listed as being located in another state. Finally, the IRS is known to do only a fair job of scrutinizing the exempt organization records and compiling the Exempt Organization Master File. Errors are consequently common in these files, reducing their reliability.

For all of these reasons, we decided not to rely exclusively on the IRS files in compiling our sampling frame but to test it against other information about nonprofit organizations in the state and supplement it as necessary. Our sampling frame was thus constructed from two broad sources: first, the IRS Business Masterfile for the Baltimore Key District Office of March 1995 (the IRS database); and second, different directories of Maryland organizations operating in fields known to involve nonprofit agencies (the directories).

The IRS database is a listing of nearly 20,000 tax exempt organizations included in the active business Master File. The Master File is reviewed monthly to purge all entities that are known to be out of business. The database thus contains a reasonably comprehensive listing of organizations with total annual income of $25,000 or more that are required to file Form 990, as well as a listing of entities with annual income under $25,000 which are not required to file the form 990 but are known to the IRS.

---

[1]This Appendix was prepared by Wojciech Sokolowski.

We filtered the IRS database to include only those charitable nonprofit entities exempt under the sections 501(c)(3) and 501(c)(4) of the Tax Code and that are not governmental units or private non-operating foundations. The filtering procedures resulted in a total of 12,981 organizations that were included to our sampling frame, of which 4,326 had income of $25,000 or more and 8,655 reported no income or assets.

To determine how complete the IRS database really was, we collected 49 directories of Maryland nonprofit entities. We then sampled a small number of records in these directories and compared them against the IRS database. We used three directories that appeared to be the most comprehensive regional sources. For the Baltimore area, we used the 1995 *Directory of Organizations in Baltimore County* published by the Baltimore County Library in November 1994. For a rural region, we used the *Washington County Human Services Providers Directory*, published by the United Way of Washington County (undated). For a suburban region, we used the *Rainbow Directory of Community Service Providers* published by the United Way of Montgomery County in 1992. We also used the *Maryland Food Bank List of Food Pantries and Soup Kitchens, Membership list of the Association of Independent Maryland Schools,* and the 1994 *Directory of Emergency and Transitional Housing Programs* published by the Maryland Department of Human Resources. To determine whether there was any need to supplement the IRS lists with the directories, we selected a random sample of 100 organizations from each directory and compared them against the IRS database.

The results varied widely by field of activity. Almost all arts and environmental organizations were also listed in the IRS database, while less than half of the private primary and secondary schools listed in the membership list of the Association of Independent Maryland Schools were included in the IRS database. This suggested that the IRS database does not include all active nonprofit organizations. Overall, we identified 14 percent more agencies in the sampled directories than were in the IRS data base.

Consequently, we decided to supplement our sampling frame with the following statewide directories: *Directory of Maryland Museums* (American Association of Museums, 1993), *Membership List of the Association of Independent Maryland Schools* (September 4, 1994), *Directory of Community Services in Maryland* (Health and Welfare Council, 1990), *Maryland Legal Service Corporation Grantee List for Fiscal Year 1995, List of Maryland Local Land Trust* (Maryland Environmental Trust, November 17, 1994), *Maryland Museums at a Glance* (Maryland Historical Trust, May 1994), *Maryland State Arts Council Fiscal Year 1995 List of Applicants, Self Advocacy Groups* (author unknown, undated), and *Senior Citizens Centers* (author unknown, undated). As reflected in Table A.1, these directories contained a cumulative total of 5,349 agencies. Altogether, therefore, we had a total of 18,330 agencies from which to draw our sample, albeit with a high likelihood that many of the agencies in the directories were also included in the IRS listings (See Table A.1).

1.2 Sample Selection

Our sampling procedure aimed at two goals. First, we wanted to attain a sufficient representation of large organizations to generate reasonable estimates of the nonprofit sector's finances. This was particularly important in view of the fact that prior research had suggested that the distribution of resources within the nonprofit sector is highly uneven, with a small number of quite large agencies generally dominating the sector's revenues and expenditures. At the same time, we wanted to include

Table A.1
Survey Sample Selection

Category	Population	Target Sample	Surveys Sent	Completed Surveys	Return Rate
IRS Working Data Base					
Large	4,326	500	2,084	318	15.3%
Small	8,655	250	986	99	10.0
Directories	5,349	N.A.	158	22	13.9
Total	18,330[a]	750	3,228	439	13.6%

N.A. = Not Applicable
[a] Includes an unknown number of duplicates assumed to be .85 percent

enough small agencies to fairly represent the full panorama of the nonprofit sector in the state, much of which is composed of small agencies. As reflected in Table A.1, to achieve a reasonable level of statistical confidence, we sought a sample of 500 organizations with income and assets and a supplementary sample of 250 small agencies. Assuming a response rate of 25 percent, this meant that we would have to aim at a survey sample of approximately 3,000 agencies. Beyond this, we wanted to ensure that the sampling process was random and that agencies not included on the IRS listings had the same chance of being included as those that were included.

Accordingly, our sampling procedure consisted of three basic steps:

First: To select our sample of 2,000 "large" agencies (i.e. agencies with income or assets in excess of $25,000) we created an imaginary data file consisting of the IRS Working Data Base and the combination of all the directories. Since we had no basis for knowing the size of the directory agencies, and since it was impractical to compare all directory entries against all IRS entries prior to sample selection in order to eliminate duplicates, we proceeded in the following fashion:

(1) With close to 10,000 agencies in the IRS large sample and the directories and a target sample size of 2,000, we decided to select every fifth agency randomly from the Working Data Base and the directories, eliminating any directory agencies selected that were also included on the IRS data base and replacing it with the next fifth entry. This was done to ensure that organizations that appeared in both the IRS data base and the directories did not have a higher chance of being selected.

(2) Once a first pass was made through the directories using this procedure, the remaining sample was filled in from the IRS data base using the same selection procedure (every fifth entry).

(3) Because the number of agencies in the three highest income brackets we targeted was extremely limited such that our likely final sample of agencies in these categories would be too small to sustain statistically valid conclusions, we decided to oversample these categories by including all organizations in these income groups in the survey sample.

Second: To select our sample of 1,000 small agencies, we relied exclusively on the IRS listings and selected every ninth agency from the

IRS data base (1,000/8,655) beginning from a randomly selected starting point.

Third: We examined the resulting survey sample and eliminated any remaining duplicates, any organizations known to be defunct, or any organizations that escaped the earlier screening and did not fit the study parameters (e.g., foundations that had been misclassified in the IRS data base).

The overall result, as Table A.1 shows, was a survey sample of 3,228 agencies, of which 2,084 were IRS database agencies thought to have income and assets in excess of $25,000, 986 were IRS database agencies thought to have income and assets less than $25,000, and 158 were organizations with unknown income and assets identified through the directories.

A survey form was developed based on the earlier *More Than Just Charity* survey to permit some comparability but augmented with questions about ethics and accountability. The surveys were mailed out in December 1995 by the Maryland Association of Nonprofit Organizations. This was followed by a series of letters and, in some cases, phone calls. Ultimately, by late March, we had received back 439 responses, for a response rate of 13.6 percent, somewhat lower than we had hoped, but still quite adequate for our purposes.

2. AN ANALYSIS OF SAMPLE REPRESENTATIVENESS

Given the lower than expected response rate to our survey, we conducted a series of tests to determine whether the low response ratio compromised the representativeness of the originally selected sample. For this purpose, we divided the originally selected sample of 3,228 entities into two groups: one consisting of agencies that did respond to the survey and the other consisting of those that did not respond. We then compared these two groups in terms of characteristics that we could examine based on the IRS database. These included agency size (measured by assets and income), geographical location (by county), and the field of activity.

Essentially, we computed the probability that these two samples came from two different populations. By convention, if this probability (the p-value) is less than 5 percent, we could assume that the samples came from different distributions. If the p-value was 5 percent or more, we could assume that the two sub-samples came from the same population.

2.1. Size

In the first test, we aimed to determine whether the two groups differed significantly in terms of the size of agencies they contained. In operational terms, we tested whether the two groups differ from each other with respect to mean income and assets, using an independent sample t-test.

The two t-tests, conducted separately for the value of assets and for income, showed no significant differences between the two groups. The results are quite robust ($p>.6$) on the 2-tailed test, meaning that there is over a 60 percent probability that the observed differences between these two groups obtained as a result of chance rather than as a result of any systematic bias in the selection of the sample. Based on these results, we concluded that the "non-response" group is not different from the "response" group in terms of the organizations' size.

2.2. Geographical Location

The second test used a similar approach to determine whether the two groups differ significantly in terms of the geographical location of agencies they contain. We constructed a contingency table, cross-tabulating the organizations by the response status (response vs. no-response) and the county code. We then conducted a Pearson Chi-square test to evaluate the probability that the frequencies observed in the contingency table differed significantly from the frequencies that could have been expected by chance, i.e. if the response status and geographical location were independent of each other. The results were marginally significant (p=0.05). This means that there is only a 5 percent chance that the difference in geographical composition between the two groups obtained by chance rather than as a result of some systematic bias.

However, a fairly large number of cells had expected frequencies lower than 5 percent, making the results at least potentially problematic. Consequently, we conducted a second test in which all counties with at least one cell with an expected frequency less than 5 were collapsed into a single category. The results of that second test yielded a p-value of <0.05, which suggests that there are significant differences between these two groups in terms of the geographical location of the included organizations.

In particular, the organizations from counties with small populations of organizations are over-represented in the "responses" group by a ratio 1.46:1 (organizations from these counties constitute 7% of the "response" group, but only 4.8% of the "non-response" group). Other counties that are slightly over-represented among the respondents are: Carroll (3% vs. 1.4%), and possibly Harford (3.6% to 2.3%) and Talbot (1.5% to 0.9%). The counties that are under-represented among the respondents by a ratio close to 1:2 are: Allegany (1.1% to 2.1%), and Charles (0.8% to 1.6%). However, given the rather small populations of these counties, those differences probably do not have much weight vis a vis the entire sample. Slightly more problematic is the under-representation of Howard County (3.2% vs. 5.2% or 1:1.6) because of a relatively large size of its organizational population (156 cases in the entire sample). On the other hand, the differences between the two sub-samples in counties with large populations of agencies are rather small.

Based on these results, we concluded that the sample of 439 agencies that returned filled out questionnaires differed from the originally selected sample of 3,228 entities in terms of their geographic composition, although the differences are primarily among the counties with small organizational universes and are likely to cancel each other out. Consequently, the effects of geographical bias on the entire sample are probably small, if not negligible.

2.3. Field of Activity

A third test compared the responding and non-responding agencies in terms of field of activity. Since the IRS uses a fairly detailed classification of the activity areas, it was necessary to collapse some of the categories in order to avoid having large numbers of categories with small numbers of observation in each category, which could easily compromise the results of the Pearson Chi square test. In particular, we created the following five groups:

- health care (IRS codes 150-179);
- inner city and community development (IRS codes 400-429);
- IRS code 590 (not listed in the code book, but containing 227 organizations);
- organizations serving other organizations (IRS codes 600 - 603); and
- all other organizations (all remaining IRS codes).

The Chi square test shows no differences between respondents and non-respondents in terms of the distribution among these aggregated areas of activity. In fact, these two groups are virtually identical in terms of areas of activity represented.

2.4 Conclusions

Based on these results we concluded that the low response rate did not substantially alter the contours of the sample of 3,228 organizations we initially selected. Measured in terms of size and areas of activity, the organizations in the survey sample and the organizations in the target sample do not differ from each other significantly. Some counties with small organizational populations seem to be slightly over-represented, while other smaller counties, such as Howard, Charles and Allegany, are under-represented. This problem does not seem to affect counties with large organizational populations, however. We could thus reasonably believe that the possible biases resulting from those under- and over-representations among smaller counties cancel-each other out and have little effect on the entire sample. Consequently, the sample of organizations that did respond can be considered to be representative of the original sample selected for the study.

3. HANDLING OF MISSING OR INCONSISTENT DATA

Given the relatively small size of the sample, however, missing or ambiguous responses to particular questions posed a nontrivial problem to data analysis because they resulted in further reduction of the effective sample size. Therefore, for the key variables we used for data analysis we had to develop routines for estimating missing values with best available estimates. In some cases, this could be done relatively easily from other information supplied on the survey form. For example, missing responses could be replaced with zeroes where other information made it clear that this was the appropriate response. For example, in cases reporting no employment and skipping all employment-related questions, we recoded missing values on variables pertaining to the number of employees as numerical zeroes. In other cases, more complex estimating routines had to be developed. These routines are described for each variable separately below (sections 3.1 to 3.5 below).

3.1. The Primary Field of Activity

As noted in the text, all agencies were assigned to a "primary service field" based on the percent of total operating expenditures spent in 14 major service areas identified on the survey (question 22). Agencies were assigned to the field in which they reported spending 51 percent or more of their total operating expenditures. Organizations that did not report spending 51 percent or more of their operating expenditures in one service area were coded as "multi-field."

There were 88 cases, however, with insufficient information to assign to a field (e.g. the sum of expenditure shares for all fields of activity was less than 50 percent). Where possible, the field of activity was determined based on a visual examination of the organization's reported program activities as well as the organization's name. Using this technique, we were able to classify all but one agency by their major fields of activity.

3.2. Organization Size

We classified all organizations in the sample according to their size based on the income reported in the survey. We used three categories for this purpose: small (less than $25,000 in income), medium ($25,000 to $999,999 in income) and large (income of $1 million or more). Insufficient information posed a considerable problem in determining the organizations' size: 86 respondents did not report any income amount for FY 1994, and of these, only 26 answered "no" to the survey question inquiring whether they had any income in either FY 1994 or FY 1993; in addition, 28 respondents did not provide any information at all on their income.

To solve this problem, we assigned 32 respondents who did not report any income amount in the survey and answered "no" to the question of whether they had any income in FY 1994 or FY 1993 to the "small" category. In those cases that did not report any income amount for FY 1994, and did not answer "no" to the question inquiring whether they had any income in FY 1994 or FY 1993, we used the income reported for agencies in the IRS database as the criterion for estimating the size category.

These steps left 13 agencies with missing values on the size variable. As it turned out, all of these cases were selected to the sample from the directories (see section on sampling methodology). We concluded that these missing cases most likely represent small agencies with income under $25,000 (larger agencies would most likely show up in the IRS database as they are required to file form 990), and classified them accordingly.

3.3. 1994 Expenditures

Similar problems arose in developing estimates for agency expenditures, which 89 respondents failed to provide. These agencies were dealt with in a similar fashion. Thus, 27 organizations reported no expenditures but answered "no" on the question of whether they had any operating expenditures for fiscal years 1994 or 1993. We therefore assumed that the value of their FY 1994 expenditures was zero. This left 62 cases with missing data on the expenditure variable for FY 1994. For 27 of these, we were able to utilize responses to two other survey questions that sought information on various types of expenditures (wages, fringe benefits, and non-personnel expenditures), leaving 35. Four other cases reported no employment or income in the IRS database and were therefore assumed to have zero expenditures in 1994.

The remaining cases all reported having expenditures in FY 1993 or 1994 but did not report any figures. For five of these, we used the income figure that the organizations reported on the assumption that in nonprofit operations income comes close to equaling expenditures since no profits are expected. For another fifteen, we were able to estimate the expenditures based on the income information reported in the IRS database. Finally, in all cases with still missing data on FY 1994 expenditures and reporting nonzero full-time equivalent[2] employment in 1994, we estimated those expenditures as including at least the cost of labor. We calculated the average labor cost based on information provided by the respondents reporting both nonzero labor cost (wages and fringe benefits) and nonzero employment. The weighted average labor cost per one full time equivalent employee thus calculated was $40,700. We then multiplied that figure by the full-time equivalent employment in the remaining cases with missing data on FY 1994 expenditures to estimate those expenditures. This procedure reduced the number of observations for which we were missing FY 1994 income information to seven.

3.4 1994 Revenue Sources

We asked the respondents to break down their total revenues among the various revenue sources (e.g., private giving, government support, fees and charges). In a small number of cases this created the problem that the reported totals did not equal the sum of the component parts. This section describes our procedure for handling such discrepancies.

In 25 cases, the subtotal of revenues from private giving sources (questions 25a to 25e) was listed by respondents as a separate variable. In 11 of these cases, both the subtotal variable and at least one of the private sources had nonzero entries. In four of these cases, the sum of the private sources and the value reported as the subtotal were equal, but in the remaining nine cases, the reported subtotals were greater than the sum of revenues from the respective

---

[2] We calculated the FTE employment figure based on information provided by the respondents.

sources. Furthermore, there were 14 cases where only the subtotal was reported, while the revenues from the specific private sources were not listed. There was no case of the sum total being greater than the reported subtotal.

We handled this problem by creating a new variable reflecting the total FY 1994 revenues from private giving sources by adding the values of revenue reported for the individual sources, and then selecting the greater of the sum total thus calculated, or the subtotal reported by the respondents. As a result of this transformation, the private subtotal we calculated from the available data is greater than the sum of all private sources reported by the respondents by $6,406,729.

We subsequently compared the total FY 1994 revenues reported by the respondents with the sum total of all reported revenue sources. In 203 cases, the total reported income equaled the sum of the sources, but in 132 cases (excluding missing data) there were discrepancies, some of them quite substantial. Specifically, we identified 76 cases with reported income greater than the sum of the sources, and 56 cases with reported income smaller than the sum of the sources. Among the former, there were 15 cases of agencies reporting a non-zero income in FY 1994, but not reporting income from any of the sources.

If the reported total is greater than the sum of the component parts, a certain portion of the known revenues could not be allocated to any specific source. For our sample, the total discrepancy between the reported total revenue and the component parts amounted to nearly $283 million, or about 15% of the total known revenues. This would not be a serious problem if that unallocated amount is more or less equally distributed across all revenue categories. However, if the discrepancy originated from underreporting of revenues from only a few sources, this could lead to a distorted image of the revenue structure of the Maryland nonprofit sector.

Given the seriousness of this problem, we decided to recontact the agencies with non-trivial discrepancies between reported totals and the sum of the reported components of revenue. Through this process, we were able to reduce the discrepancy to less than 1 percent of total revenue.

3.5. Volunteers

The amount of volunteer work was reported as the total number of volunteer hours used by the agency during the past 12 months, and the number of volunteers that worked at the agency during that period of time. In cases where neither the number of volunteers, nor the number of volunteer hours was provided, we assumed that the agency does not use volunteers, and transformed the values on both variables to numerical zeroes. Where only one type of information was provided, while the information on the other variable was missing, we declared the missing information as "user missing."

4. Weights

To transform our sample of 439 respondents into a statistically reliable representation of the entire population of Maryland nonprofit agencies, it was necessary to calculate weights to blow the sample up to the population taking account of the differences between the composition of the sample and the composition of the entire population of agencies along the dimensions we could measure.

For this purpose, we used the 12,981 observations in the IRS database as the most reliable indicator of the dimensions of the population available. More specifically, we divided the sample and the sampling frame into three groups: agencies with income under $25,000, agencies with income between $25,000 and $1 million, and agencies with income over $1 million. We then calculated the probability of being selected into each group by dividing the respective numbers of records in the sample by the numbers of records in the IRS sampling frame. The weights used to blow the sample observations up to the population were then the reciprocals of the probabilities of being selected, or 1/(probability of being selected). For this purpose, the 13 cases with insufficient income data (see section 2.1 above) were assigned to the category "under $25,000," which produced a count of 93 cases in that category. As a result, 93 organizations with income under $25,000 received the weight of 93.065. The 239 organizations with income between $25,000 and $1 million received a weight of 14.542. And 107 organizations with income over $1 million were assigned the weight of 8.009. Thus, when the entire sample is weighted, the total number of "cases" equals 12,981, the number of cases in the sampling frame derived from the IRS database.

We used these weighted results in univariate analyses to estimate the overall size and scope of the nonprofit sector in Maryland (e.g., expenditures, revenues, employment, volunteer work, and the relative size of different fields of service).

However, multivariate analyses (contingency tables) involving measures of statistical significance (e.g. Pearson Chi square) depend on the actual number of cases in a sample. Since the weights we employed blow up that number quite substantially, the significance tests calculated on that basis would be simply misleading. To bypass this difficulty and still be able to conduct multivariate analyses reflecting differences in probability of being selected for different groups of agencies, we created an alternative system of weights that do not change the number of cases in the sample. We divided the weights we used for univariate analyses by a constant representing the ratio of the number of organizations in the sampling frame to the number of organizations in the sample (12,981/439=29.569). We used this alternative system of weights exclusively for calculating statistical significance levels. To report the distribution of cases in contingency tables, however, we used the system of weights that estimates the actual number of agencies in the population.

APPENDIX B
FIELDS OF ACTIVITY OF MARYLAND NONPROFIT AGENCIES

N=12,981

CODE	SERVICE FIELD	Agencies Reporting Activity in Field Number	Percent
Social Services		6,662	51.3%
A001	Adoption	183	1.4%
A002	Foster Care	193	1.5%
A003	Child Care/Day Care	775	6.0%
A004	Child Protective Services	217	1.7%
A005	Other Child Welfare	754	5.8%
A006	Youth Clubs & Associations	1,214	9.3%
A007	Family Life, Parent Education, Parenting Skills	1,359	10.5%
A008	Family Violence Shelters and Services	742	5.7%
A009	Homemaker/Chore/In-home Support Services	461	3.6%
A010	Individual and Family Counseling	1,280	9.9%
A011	Peer Counseling & Self Help	1,113	8.6%
A012	Emergency Assistance (Food, Money, Clothing)	2,330	18.0%
A013	Direct Financial Assistance	1,608	12.4%
A014	Transportation - Free or Subsidized	1,263	9.7%
A015	Homeless Persons Centers & Services	945	7.3%
A016	Disaster Preparedness & Relief	507	3.9%
A017	Free Food Distribution	1,235	9.5%
A018	Food Bank/Food Pantry	1,052	8.1%
A019	Nutrition/Meal Services	822	6.3%
A020	Institutional Care	379	2.9%
A021	Short Term Residential Care	273	2.1%
A022	Group Home	721	5.6%
A023	Hospice	320	2.5%
A024	Senior Center	570	4.4%
A025	Services for Disabled	618	4.8%
A026	Thrift Shop	320	2.5%
A027	Public Safety Education	1,016	7.8%
A028	Information & Referral	1,926	14.8%
A029	Other Social Services	1,029	7.9%

Mental Health		**1,908**	**14.7%**
B030	Alcoholism & Alcohol Abuse	660	5.1%
B031	Drug/Substance Abuse, Chemical Dependency	823	6.3%
B032	Addiction/Obsessive Behavior Related	389	3.0%
B033	Psychiatric Care - General	477	3.7%
B034	Mental Health Clinic	312	2.4%
B035	Community Mental Health Center	337	2.6%
B036	Residential Treatment - Mental Health Related	281	2.2%
B037	Transitional Residential Care/Treatment -Mental Health Related	265	2.0%
B038	Hotline - Crisis Intervention	329	2.5%
B039	Suicide Prevention	225	1.7%
B040	Rape Victim Services	110	0.8%
B041	Evaluation & Testing	161	1.2%
B042	Other Mental Health Services	526	4.1%
General Health		**2,001**	**15.4%**
C043	Skilled Nursing Home	172	1.3%
C044	Intermediate Nursing/Care Facility	84	0.6%
C045	Hospital - General	292	2.3%
C046	Primary/Specialty Medical Care	432	3.3%
C047	Maternal & Child Health	345	2.7%
C048	Reproductive Health and Family Planning	108	0.8%
C049	Prenatal/Childbirth	215	1.7%
C050	Rehabilitation	255	2.0%
C051	Home Health Care	238	1.8%
C052	Health Counseling, Screening, Prevention	514	4.0%
C053	Other Health Services	969	7.5%
Education		**5,978**	**46.1%**
D054	Pre-school Education	1,128	8.7%
D055	Nursery School	720	5.5%
D056	Kindergarten	1,441	11.1%
D057	Primary/Elementary School	2,290	17.6%
D058	Secondary/High School	1,974	15.2%
D059	Higher Education	1,455	11.2%
D060	Vocational, Technical Education	470	3.6%
D061	Adult Continuing Education	1,009	7.8%
D062	Remedial Education & Tutoring	1,039	8.0%
D063	Special Education	949	7.3%
D064	Dropout Prevention	323	2.5%
D065	Parent Teacher Group	983	7.6%
D066	Library or Information Center	1,313	10.1%
D067	Other Education & Instruction	2,192	16.9%
Community Development		**3,711**	**28.6%**
E068	Housing Development & Construction	547	4.2%
E069	Housing Improvement & Repair Assistance/Rehabilitation	510	3.9%

E070	Housing Management	459	3.5%
E071	Homeless, Temporary Shelter	550	4.2%
E072	Housing Search Assistance	587	4.5%
E073	Housing Owners, Renters Organizations	361	2.8%
E074	Energy Assistance	374	2.9%
E075	Other Housing Services	616	4.7%
E076	Community/Neighborhood Development & Preservation/Improvement	1,253	9.7%
E077	Urban, Community Economic Development	383	2.9%
E078	Rural Development	90	0.7%
E079	Promotion of Business	558	4.3%
E080	Business & Economic Development	560	4.3%
E081	Community Service Clubs (e.g. Kiwanis and Lions)	979	7.5%
E082	Other Community, Economic Development	787	6.1%
Environment & Animals		**2,131**	**16.4%**
F083	Pollution Abatement & Control Services	405	3.1%
F084	Natural Resources Conservation & Protection	917	7.1%
F085	Botanical, Horticultural & Landscape Services	424	3.3%
F086	Environmental Beautification & Open Spaces	891	6.9%
F087	Environmental Education & Outdoor Survival	575	4.4%
F088	Other Environmental	747	5.8%
F089	Animal Protection & Welfare	243	1.9%
F090	Wildlife Preservation, Protection	283	2.2%
F091	Veterinary Services	58	0.4%
F092	Zoo, Zoological Society	52	0.4%
F093	Aquariums	37	0.3%
F094	Other Animal Related	89	0.7%
Employment & Job Training		**1,661**	**12.8%**
G095	Employment Procurement Assistance	526	4.1%
G096	Job Development	574	4.4%
G097	Vocational Counseling	436	3.4%
G098	Employment Training, Including Apprenticeship	595	4.6%
G099-A	Vocational Rehabilitation	466	3.6%
G100	Subsidized Employment	347	2.7%
G101	Other Employment & Training, Job Services	788	6.1%
Advocacy & Legal Services		**4,041**	**31.1%**
H102	Legal Aid & Counseling, Civil & Criminal	269	2.1%
H103	Public Interest Litigation	15	0.1%
H104	Advocacy in Mental Health	817	6.3%
H105	Advocacy in Social Services	1,259	9.7%
H106	Advocacy in Health	1,278	9.8%
H107	Advocacy in Education & Instruction	1,116	8.6%
H108	Advocacy in Housing	908	7.0%

H109	Advocacy in Community, Economic Development	648	5.0%
H110	Advocacy in Environmental Issues	371	2.9%
H111	Advocacy in Animal Related	16	0.1%
H112	Advocacy in Employment & Training, Jobs	474	3.7%
H113	Advocacy in Arts & Culture	396	3.0%
H114	Advocacy in Recreation & Sports	275	2.1%
H115	Advocacy in Criminal Justice	325	2.5%
H116	Advocacy in International Affairs	108	0.8%
H117	Elimination of Discrimination/Prejudice	835	6.4%
H118-A	Equal Opportunity & Access	498	3.8%
H118-B	Civil & Human Rights	848	6.5%
H119	Immigrant and/or Refugee Rights	58	0.4%
H120	Minority Rights	631	4.9%
H121	Disabled Person's Rights	909	7.0%
H122	Women's Rights	604	4.7%
H123	Race/Intergroup Relations	494	3.8%
H124	Voter Education/Registration	220	1.7%
H125	Other Advocacy, Legal Services Civil Rights	833	6.4%
Arts & Culture		**3,335**	**25.7%**
I126	Performing Arts Productions/Companies	1,586	12.2%
I127	Museums	769	5.9%
I128	Art Galleries	434	3.3%
I129	Visual Arts Organizations, Services	341	2.6%
I130	Media, Communications Organizations	484	3.7%
I131	Folk Arts	289	2.2%
I132	Training or Workshops in Culture or Arts	947	7.3%
I133	Arts Service Organizations and Activities	284	2.2%
I134	Arts Council, Agency	174	1.3%
I135	Humanities Organizations	238	1.8%
I136	Historical Societies & Activities	773	6.0%
I137	Cultural and Ethnic Awareness	995	7.7%
I138	Other Arts and Culture	775	6.0%
Sports & Recreation		**1,784**	**13.7%**
J139	Camping - Day & Overnight	361	2.8%
J140	Recreation Facilities (Parks, Playgrounds etc)	653	5.0%
J141	Recreational, Pleasure, or Social Clubs	481	3.7%
J142	Amateur Sports Clubs, Leagues	527	4.1%
J143	Amateur Sports Competitions	409	3.2%
J144	Sports Training Facilities, Agencies	142	1.1%
J145	Professional Athletic Leagues	15	0.1%
J146	Other Recreation, Sports, and Leisure	591	4.6%

Crime, Criminal Justice		**1,353**	**10.4%**
K147	Juveniles Justice & Delinquency Prevention	489	3.8%
K148	Transitional Care (Halfway House) for Offenders	424	3.3%
K149	Rehabilitation Services for Offenders	449	3.5%
K150	Services to Prisoners & Families, Multipurpose	273	2.1%
K151	Other Crime, Criminal Justice	474	3.7%
International		**1,206**	**9.3%**
L152	International Development, Relief Services	856	6.6%
L153	Promotion of International Understanding	413	3.2%
L154	International Peace & Security	23	0.2%
L155	Foreign Policy Research & Analysis	0	0.0%
L156	International Human Rights	137	1.1%
L157	Immigrant and/or Refugee Rights	23	0.2%
L158	Other International Affairs	365	2.8%
Philanthropy, Volunteerism		**2,516**	**19.4%**
M159	Grantmaking	609	4.7%
M160	Volunteer Recruitment, Referral, Placement	991	7.6%
M161	Volunteer. Training & Supervision	1,193	9.2%
M162	Fundraising Organizations	1,559	12.0%
Research		**2,132**	**16.4%**
N163	Science Research, General	235	1.8%
N164	Physical Sciences, Earth Sciences Research	177	1.4%
N165	Engineering & Technology Research	69	0.5%
N166	Biological, Life Science Research	156	1.2%
N167	Social Science Research	185	1.4%
N168	Research in Social Services	364	2.8%
N169	Research in Health	535	4.1%
N170	Research in Mental Health	341	2.6%
N171	Research in Education & Instruction	215	1.7%
N172	Research in Housing	231	1.8%
N173	Research in Community, Economic Development	377	2.9%
N174	Research in Employment & Training, Jobs	105	0.8%
N175	Research in Advocacy, Legal Services, Civil Rights	116	0.9%
N176	Research in Arts & Culture	455	3.5%
N177	Research in Recreation, Sports	23	0.2%
N178	Interdisciplinary Research	191	1.5%
N179	Public Policy Research & Analysis	505	3.9%
N180	Other Research	564	4.3%

ABOUT THE AUTHOR

Dr. Lester M. Salamon

Lester M. Salamon is a Professor at The Johns Hopkins University and was Founding Director of the university's Institute for Policy Studies, a post he left in July 1997 to head the Institute's Center for Civil Society Studies. Dr. Salamon previously served as Director of the Center for Governance and Management Research at The Urban Institute in Washington, D.C. and was Deputy Associate Director of the U.S. Office of Management and Budget in the Executive Office of the President. Before that, he taught at Duke University, Vanderbilt University, and, during the American civil rights struggle of the mid-1960s, at Tougaloo College in Tougaloo, Mississippi.

Dr. Salamon was a pioneer in the empirical study of the nonprofit sector. His 1982 book, *The Federal Budget and the Nonprofit Sector*, was the first to document the scale of the American nonprofit sector and the extent of government support to it. As director of the Johns Hopkins Comparative Nonprofit Sector Project, Dr. Salamon has extended this analysis to the international sphere, producing the first comparative empirical assessment ever undertaken of the size, structure, financing, and role of the nonprofit sector at the global level.

Dr. Salamon is the author or editor of more than a dozen books and has contributed articles to more than 50 different journals. His book, *America's Nonprofit Sector: A Primer*, is the standard text used in college-level courses on the nonprofit sector in the United States. His *Partners in Public Service: Government and the Nonprofit Sector in the Modern Welfare State*, published in 1995 by the Johns Hopkins University Press, won the 1996 ARNOVA Award for Distinguished Book in Nonprofit and Voluntary Action Research.

Dr. Salamon received his B.A. degree in Economics and Policy Studies from Princeton University and his Ph.D. in Government from Harvard University. He is married, has two sons, and serves on the Board of the International Society for Third-Sector Research and on the Editorial Boards of *Voluntas, Administration and Society,* and *Nonprofit and Voluntary Sector Quarterly*.

WITH ASSISTANCE FROM:

Peter V. Berns

Mr. Berns is the Executive Director of the Maryland Association of Nonprofit Organizations (Maryland Nonprofits). Prior to becoming Executive Director of Maryland Nonprofits he was the Deputy Chief of the Consumer Protection Division of the Office of the Attorney General of Maryland. Before moving to Maryland he provided legal representation to nonprofit groups in Washington, D.C.

Mr. Berns is active as a volunteer and serves on boards of the National Council of Nonprofit Associations, the Baltimore Jewish Council, and the Maryland Food Committee.

Mr. Berns received his JD, *cum laude*, from Harvard Law School and has an LLM in advocacy from Georgetown University Law Center. He received his BA in psychology *magna cum laude* from the University of Pennsylvania.

Amy Coates Madsen

Amy Coates Madsen is the Special Assistant to the Executive Director/Policy Analyst for the Maryland Association of Nonprofit Organizations. Ms. Madsen's primary responsibilities include: fiscal analysis for the public policy division of the association and coordination for the Maryland Nonprofits Ethics and Accountability in the Nonprofit Sector Initiative.

Ms. Madsen received her Master of Arts in Policy Studies degree from the Johns Hopkins University Institute for Policy Studies in Baltimore, Maryland; and her Bachelors degree in History from the Virginia Polytechnic Institute and State University in Blacksburg, Virginia. Ms. Madsen has been inducted into the Phi Beta Kappa society.

Dr. S. Wojciech Sokolowski

S. Wojciech Sokolowski, Ph.D., is a Research Associate at the Johns Hopkins University Institute for Policy Studies in Baltimore. Dr. Sokolowski also teaches at Morgan State University in Baltimore. He holds a Ph.D. in Sociology from Rutgers University (1997), a Master of Arts degree in Sociology from San Jose State University (1990), and a Master of Arts in Philosophy from Lublin Catholic University (1978).

Prior to his service at the Institute for Policy Studies and Morgan State University, Dr. Sokolowski taught at Hartnell College in Salinas, California and the Defense Language Institute in Monterey, California. He has also served as a consultant for the Santa Clara County Board of Supervisors.

In addition to his doctoral dissertation, *The Role of Professionals and Nonprofit Organizations in the Political-Economic Reform in Poland, 1989-1993,* Dr. Sokolowski has written and presented extensively in the area of international nonprofit organizations.

Dr. Stefan Toepler

Stefan Toepler, Ph.D., is a Research Associate at the Johns Hopkins University Institute for Policy Studies in Baltimore. He holds a Ph.D. in Economic Studies (1995) and a graduate degree in Management (1991) from the Department of Economic and Administrative Studies at the Free University Berlin in Germany.

From 1992 to 1994, he held a German Research Society scholarship at the Free University's John F. Kennedy Institute for North American Studies. During the 1993/1994 academic year, he was also a philanthropy fellow at the Johns Hopkins Institute for Policy Studies. He has published three books in German and numerous articles and book chapters in English and German on issues ranging from financing the arts to the comparative study of philanthropic foundations.